Elevate SwiftUI Skills by Building Projects

Build four modern applications using Swift, Xcode 14, and SwiftUI for iPhone, iPad, Mac, and Apple Watch

Frahaan Hussain

<packt>

BIRMINGHAM—MUMBAI

Elevate SwiftUI Skills by Building Projects

Group Product Manager: Rohit Rajkumar

Publishing Product Manager: Nitin Nainani

Senior Editor: Aamir Ahmed

Technical Editor: Simran Udasi

Copy Editor: Safis Editing

Project Coordinator: Aishwarya Mohan

Proofreader: Safis Editing

Indexer: Tejal Daruwale Soni

Production Designer: Shyam Sundar Korumilli

DevRel Marketing Coordinator: Nivedita Pandey

First published: August 2023

Production reference: 1090823

Published by Packt Publishing Ltd.
Grosvenor House
11 St Paul's Square
Birmingham
B3 1RB, UK.

ISBN 978-1-80324-207-1

www.packtpub.com

To my amazing wife Maimuna,

Thank you for your unwavering support throughout my book-writing journey. You've been my rock, my muse, and my strength. Your faith in me and your understanding of the sacrifices required have propelled me forward. I dedicate this book to you, my partner and inspiration.

To my beloved Arya and Aman,

You've brought immense joy to our lives. I hope you embrace programming like me, but also explore other passions. We'll support you every step of the way. May you inspire each other and build a strong bond. Remember, your worth extends beyond your chosen path. I love you and can't wait to see what you'll achieve.

– Frahaan Hussain

Foreword

I take great pleasure and admiration in presenting to you the remarkable accomplishments of Frahaan Hussain. In the following pages, you will embark on a journey that showcases the exceptional expertise and unwavering commitment of an individual who has made a profound impact on the world of programming and education.

Frahaan's achievements speak volumes. As the author of three insightful books, he has generously shared his wealth of knowledge and experience with countless readers, guiding them on a path of discovery and personal growth. His remarkable ability to simplify complex concepts and present them in an engaging manner is a testament to his outstanding communication skills and deep understanding of the subject matter.

However, Frahaan's contributions extend far beyond the written word. Through his online courses, he has empowered over half a million students, equipping them with the skills and confidence to excel in the ever-evolving field of technology. The immeasurable impact he has made on their lives has earned him the trust and admiration of aspiring programmers worldwide, who view him as a trusted mentor and a source of inspiration.

With a dedicated following of over 40,000 subscribers on YouTube, Frahaan has built a thriving community of learners who eagerly anticipate his insights and expertise. Through his engaging and informative videos, he has fostered a sense of connection and camaraderie, creating a platform for discussions and the exchange of ideas.

What distinguishes Frahaan is not only his mastery of the subject matter but also his unwavering passion for lifelong learning. His academic accomplishments, including graduating with honors in computer games programming from De Montfort University, underscore his commitment to excellence. His academic success was only the beginning, as he quickly ascended to the position of module leader at his alma mater, a prestigious role reserved for the most exceptional individuals.

Furthermore, Frahaan's consultancy work with industry giants such as Google and Chukong highlights his ability to apply his expertise in real-world scenarios, bridging the gap between theory and practice. His collaborations with these renowned clients speak volumes about his proficiency and the trust placed in his abilities.

Yet, amid his own remarkable achievements and accolades, Frahaan remains grounded and dedicated to the growth and development of others. He recognizes the transformative power of education and actively seeks to make a difference in the lives of those around him. Through his teaching, mentoring, and relentless pursuit of knowledge, he strives to ignite the spark of curiosity in others, empowering them to embrace the limitless possibilities within the realm of programming.

As you delve into the pages of this book, prepare to be inspired, enlightened, and challenged. Frahaan's words will guide you through a world of programming, offering insights, strategies, and practical advice that will sharpen your skills and deepen your understanding. Whether you are a seasoned professional or a curious beginner, the wisdom contained within these chapters will undoubtedly expand your horizons and unlock new avenues of success.

It is an honor to join Frahaan Hussain on this remarkable journey, as he exemplifies the true spirit of innovation, education, and progress. Brace yourself for an adventure that will transform the way you think, create, and navigate the world of programming.

Sunny Pradhan

Director, Vision Hive

Contributors

About the author

Frahaan Hussain is a highly accomplished individual and has established himself as a successful author with three published books. His online courses have attracted an impressive enrollment of over 500,000 students, and he enjoys the support of a dedicated YouTube following of more than 47,000 subscribers. Prior to embarking on his journey of online teaching, Frahaan excelled in his academic pursuits and graduated with honors in computer games programming from De Montfort University, ranking at the top of his class. Recognizing his expertise and potential, he was invited back to his alma mater to serve as a module leader within just two years. Alongside his consultancy work with prominent clients such as Google and Chukong, Frahaan remains dedicated to expanding the knowledge and skills of others, while continuously enhancing his own education.

About the reviewer

Kiran Jasvanee is a lead software engineer for smartSense Consulting Solutions, where he heads up the mobile app team.

Before joining smartSense, Kiran worked as a senior software engineer for TatvaSoft and OpenXcell. With over a decade of experience working in the software industry, he has worked on cross-platform mobile app development technologies: Flutter and React Native. His passion for iOS technology also led him to contribute to the open source community on GitHub. He is also an artist and completed his postgraduate degree in computer science at Bangalore University.

Kiran thanks his mentor, Mr. Mayur Pabari, who gave him the motivation and drive to be better.

Table of Contents

6

Mac Project – App Store Bars   155

7

Mac Project – App Store Body   173

8

Watch Project – Fitness Companion Design   191

Preface

Welcome to *Elevate SwiftUI Skills by Building Projects*. This book is your ultimate companion for mastering the art of building projects using Swift and SwiftUI, two powerful technologies at the forefront of Apple platform development. Whether you're just starting your journey or already have experience, this comprehensive guide is tailored to help you take your skills to the next level.

Swift and SwiftUI have revolutionized the way developers create applications for iPhone, iPad, Mac, and Apple Watch. With their intuitive syntax, robust features, and seamless integration, they offer endless possibilities for crafting stunning and performant user interfaces.

In this book, we'll dive deep into the core concepts of Swift and SwiftUI, unraveling their intricacies and exploring their full potential. Through a series of hands-on projects, you'll gain practical experience and a deeper understanding of how to leverage these technologies to build real-world applications.

Whether you're interested in creating dynamic user interfaces, integrating with backend services, implementing animations and transitions, or even exploring advanced topics such as data persistence and accessibility, this book has you covered. Each project is carefully crafted to tackle a specific aspect of app development, allowing you to learn and apply new techniques along the way.

As you progress through the chapters, you'll not only enhance your coding skills but also learn best practices and design patterns that will enable you to write clean, maintainable, and scalable code. You'll discover the power of SwiftUI's declarative syntax, its powerful data-binding capabilities, and its seamless integration with other Apple frameworks.

To facilitate your learning, each project is accompanied by detailed explanations, code samples, and step-by-step instructions. You'll also find tips, tricks, and insights from experienced developers to help you overcome common challenges and make the most out of SwiftUI's rich ecosystem.

By the end of this book, you'll have a strong foundation in Swift and SwiftUI, as well as the confidence to tackle your own projects. You'll be equipped with the skills to build impressive, user-friendly applications that leverage the full potential of Apple's platforms.

So, whether you're a beginner eager to enter the exciting world of app development or an experienced developer seeking to expand your knowledge, *Elevate SwiftUI Skills by Building Projects* will be your go-to resource for mastering Swift and SwiftUI and building amazing applications. Get ready to elevate your skills and embark on an exciting journey of creativity and innovation. Let's dive in!

As you progress through the chapters of this book, you will gain a deeper understanding of Swift and SwiftUI and their application across various Apple platforms. Each chapter is structured to provide step-by-step guidance and hands-on experience, empowering you to build real-world projects and become a proficient developer in the Apple ecosystem. Let's embark on this exciting journey together and master the art of Swift and SwiftUI development.

Who this book is for

This book is intended for aspiring iOS, iPadOS, macOS, and watchOS developers who have a basic understanding of programming concepts and a working understanding of Swift. It is also suitable for experienced developers looking to transition to SwiftUI and explore building projects across multiple Apple platforms. Whether you are a student, a professional, or a hobbyist, this book will serve as a valuable resource to expand your skills and unlock the potential of Swift and SwiftUI.

What this book covers

Chapter 1, *Swift and SwiftUI Recap*, reviews the fundamentals of the Swift programming language and the SwiftUI framework. It will serve as a refresher for those familiar with the concepts and will provide a solid foundation for beginners.

Chapter 2, *iPhone Project – Tax Calculator Design*, explores the design and layout of a tax calculator app for iPhone. Learn about UI elements and user interaction patterns to create an intuitive and visually appealing user interface.

Chapter 3, *iPhone Project – Tax Calculator Functionality*, shows how to implement the functionality of the tax calculator app for iPhone, handle user inputs, perform calculations, and display accurate results to provide a seamless user experience.

Chapter 4, *iPad Project – Photo Gallery Overview*, provides an overview of the photo gallery app for iPad. Discover the key features and user interface components necessary to create an engaging and immersive photo viewing experience.

Chapter 5, *iPad Project – Photo Gallery Enhanced View*, shows how to enhance the photo gallery app for iPad with additional features and advanced user interface elements, implement navigation patterns, and incorporate rich interactions to elevate the user experience.

Chapter 6, *Mac Project – App Store Bars*, shows how to design the navigation and toolbar elements for an app store app on the Mac platform. You will learn about layout options, customization, and creating a consistent user interface across Mac applications.

Chapter 7, *Mac Project – App Store Body*, covers building the main content area of the app store app for Mac. You will learn how to implement search functionality, display app listings, and manage user reviews to create a seamless and engaging user experience.

Chapter 8, *Watch Project – Fitness Companion Design*, shows how to design the user interface of a fitness companion app for Apple Watch. Optimize for the small screen and leverage touch interactions to create a compelling and intuitive experience.

Chapter 9, *Watch Project – Fitness Companion UI*, implements the user interface components and functionality of the fitness companion app for Apple Watch. You will learn how to track and display fitness metrics, provide notifications, and support workout sessions to enhance the user's fitness journey.

To get the most out of this book

To get the most out of this book, it is assumed that you have a basic understanding of programming concepts and a working understanding of Swift and SwiftUI. Prior experience with iOS, iPadOS, macOS, or watchOS development would be beneficial but is not required. Additionally, you should have access to a computer with macOS, as the book primarily focuses on development for Apple platforms.

The software and hardware covered in this book include the following:

- **Software**: Xcode 12 (or later) – the **integrated development environment (IDE)** for Apple platforms

- **Operating system**: macOS 10.15 Catalina (or later) – required for Xcode installation and development

- Swift 5.0 (or later) – the programming language used for developing applications on Apple platforms

- SwiftUI – the framework for building user interfaces across Apple platforms

Please ensure that you have the necessary software and hardware available to follow along with the examples and exercises in this book.

In terms of additional installation instructions, it is recommended to have Xcode installed on your macOS system. You can download and install Xcode from the Mac App Store or the Apple Developer website. If you encounter any issues during the installation process, please refer to the official documentation provided by Apple.

For readers using the digital version of this book, we recommend typing the code examples yourself rather than copying and pasting them. This will help you better understand and internalize the concepts being demonstrated, as well as minimize the chances of encountering any errors due to formatting or syntax differences.

Enjoy your journey of learning and building projects with Swift and SwiftUI! If you have any questions or need further assistance, please refer to the resources provided in the book or reach out to the author or publisher. Happy coding!

Download the example code files

You can download the example code files for this book from GitHub at `https://github.com/PacktPublishing/Elevate-SwiftUI-Skills-by-Building-Projects`. If there's an update to the code, it will be updated in the GitHub repository.

We also have other code bundles from our rich catalog of books and videos available at `https://github.com/PacktPublishing/`. Check them out!

Conventions used

There are a number of text conventions used throughout this book.

`Code in text`: Indicates code words in text, database table names, folder names, filenames, file extensions, pathnames, dummy URLs, user input, and Twitter handles. Here is an example: "In this example, we create `ColorCircle`, a custom view that conforms to the Animatable protocol."

A block of code is set as follows:

```
if ( MAX_COUNT == counter )
{
print( "Counter is 10" )
print( "Well done" )
}
```

Bold: Indicates a new term, an important word, or words that you see onscreen. For instance, words in menus or dialog boxes appear in **bold**. Here is an example: "Select **System info** from the **Administration** panel."

> **Tips or important notes**
> Appear like this.

Get in touch

Feedback from our readers is always welcome.

General feedback: If you have questions about any aspect of this book, email us at `customercare@packtpub.com` and mention the book title in the subject of your message.

Errata: Although we have taken every care to ensure the accuracy of our content, mistakes do happen. If you have found a mistake in this book, we would be grateful if you would report this to us. Please visit `www.packtpub.com/support/errata` and fill in the form.

Piracy: If you come across any illegal copies of our works in any form on the internet, we would be grateful if you would provide us with the location address or website name. Please contact us at `copyright@packt.com` with a link to the material.

If you are interested in becoming an author: If there is a topic that you have expertise in and you are interested in either writing or contributing to a book, please visit `authors.packtpub.com`.

Share Your Thoughts

Once you've read *Elevate SwiftUI Skills by Building Projects*, we'd love to hear your thoughts! Scan the QR code below to go straight to the Amazon review page for this book and share your feedback.

https://packt.link/r/1-803-24207-8

Your review is important to us and the tech community and will help us make sure we're delivering excellent quality content.

Download a free PDF copy of this book

Thanks for purchasing this book!

Do you like to read on the go but are unable to carry your print books everywhere? Is your eBook purchase not compatible with the device of your choice?

Don't worry, now with every Packt book you get a DRM-free PDF version of that book at no cost.

Read anywhere, any place, on any device. Search, copy, and paste code from your favorite technical books directly into your application.

The perks don't stop there, you can get exclusive access to discounts, newsletters, and great free content in your inbox daily

Follow these simple steps to get the benefits:

1. Scan the QR code or visit the link below

https://packt.link/free-ebook/9781803242071

2. Submit your proof of purchase

3. That's it! We'll send your free PDF and other benefits to your email directly

1
Swift and SwiftUI Recap

Firstly, I would like to thank you for reading my book, be it bought or borrowed, or whether you're having a sneak peek in the Amazon preview, I thank you.

This chapter will recap Swift and SwiftUI. We will first cover the coding standards used throughout the book for our upcoming projects and the history of Swift and SwiftUI. Then, we will take a look at the requirements for going through the projects in this book. Coding standards can be very polarizing for programmers, but they really shouldn't be. If there are any you disagree with, feel free to tweet me at @SonarSystems and let me know why. But don't let that detract from the book and what you can get from it.

Afterward, we will look at some specific SwiftUI code examples along with previews to close off the recap. We will look at how we can use views and controls; these are the visual building blocks of your application's user interface. We will use them throughout the book to draw and organize our application's content on screen. Next, we will look at layouts and presentations to learn how we can combine views in stacks, create groups and lists of views dynamically, and define view presentations and hierarchy. Hope you enjoy the chapter!

If you have any questions, feel free to join my Discord: https://discord.gg/7e78FxrgqH.

In this chapter, we will cover the following topics:

- What is Swift?
- What is SwiftUI?
- Understanding and implementing views
- Understanding and implementing layouts

By the end of this chapter, you will have learned the history of Swift and SwiftUI, and how to implement basic components from SwiftUI; this will serve as the foundation for the projects we will create in this book.

Technical requirements and standards

This book requires you to download Xcode version 14 or above from Apple's App Store.

To install Xcode, just search for Xcode in the App Store and select and download the latest version. Open Xcode and follow any additional installation instructions. Once Xcode has opened and launched, you're ready to go.

Version 14 of Xcode has the following features/requirements:

- Includes SDKs for iOS 16, iPadOS 16, macOS 12.3, tvOS 16, and watchOS 9.
- Supports on-device debugging in iOS 11 or later, tvOS 11 or later, and watchOS 4 or later.
- Requires a Mac running macOS Monterey 12.5 or later.

Download the sample code from the following GitHub link:

```
https://github.com/PacktPublishing/Elevate-SwiftUI-Skills-by-Building-
Projects
```

Here are the hardware requirements:

- You need an Intel or Apple Silicon Mac
- 4GB RAM or more

Here are the software requirements:

- macOS 11.3 (Big Sur or later)
- Xcode 14
- iOS 16 for iPad/iPhone real device testing
- watchOS 9.0 for Watch real device testing
- tvOS 16.0 for Apple TV real device testing

Here are some extra requirements:

- Intermediate knowledge of Swift
- Intermediate knowledge of another object-oriented programming language such as C++ or Objective-C

> **Important note**
> Though you can test the applications in the simulator that Xcode provides, it is highly recommended to test them on real devices.

Standards used

In this section, we will look at the coding standards that are used throughout this book. It is important to have consistent standards and know what the standards are.

Why do we need coding standards?

It is important to write good code and good code isn't just code that runs well but code that is easily maintainable and readable. Good code is an art form.

In the following sections, we will go through a set of standards that are used in the Swift programming language and these will be used throughout this book. If you do not fully agree with the standards, that is fine, but I felt it important to list the standards used in case you come across something you have never seen before, such as Yoda conditions – do any of you use them? If so, tweet me at @SonarSystems.

Indentation

You should always indent your code to be aligned with other code in the same hierarchy level. Use real tabs instead of spaces for indenting code. You can see this in the following code snippet:

```
if ( 10 == counter )
{
    print( "Counter is 10" )
    print( "Well done" )
}
else
{ print( "Wrong" ) }
```

This is helpful because you can easily see where in the hierarchy the code belongs. Xcode provides a handy little shortcut for indenting code; simply press ^ + *I* on the keyboard.

Brace style

You should always use Allman braces (named after *Eric Allman*) when writing code for structures, even if it is only one line (one line doesn't require braces in many languages, such as C++, but they should be used for ease of readability). If you only have one line of code in the structure, put the braces and code on one line. I prefer not to put the opening brace on the structure's first line. The only situation in which I would put the braces on the same line would be if there was no code in the structure yet; then, put the opening and closing brace on the same line with a space.

The following code snippet shows the preferred brace style:

```
if ( 10 == counter )
{
    print( "Counter is 10" )
```

```
    print ( "Well done" )
  }
  else
  { print ( "Wrong" ) }
```

This is helpful because it helps maintain the code and helps with readability when trying to figure out where the structure starts and ends. This is especially useful when going through loads of files really fast and trying to figure out problems.

Space usage

The use of spaces and lines varies depending on what you are using in your code. Remove any trailing whitespace (whitespace at the end of the line). You can see this in the following code snippet:

```
  print ( "Well done" )
```

Some editors automatically remove trailing whitespace, but some don't. This can cause merge conflicts. You should coordinate with the people you're working with (colleagues and the open source crowd) and what strategy everybody is using and make sure you all use the same strategy.

Comma and colon usage

When using commas/colons, put a single space after the comma/colon. You can see this in the following example:

```
  func Greet ( person: String, alreadyGreeted: Bool ) -> String
```

This makes it easier to read.

Spaces in parenthesis

Put a single space on either side of both the opening and closing parentheses for if, else, else if, for, while, and other control structures (some languages may have other control structures, so just apply these standards to those structures). You can see this in the following code snippet:

```
  if ( 10 == counter )
```

These spaces make it easier to understand, especially at a glance.

Unary operators

When using unary operators such as ++ and – in a statement, put a single space after them. You can see this in the following snippet:

```
  for ( i = 0; i < 5; i++ )
```

The spaces keep it consistent with the rest of the code and make it easier to glance over.

Parentheses spaces for functions

When defining a function, put a single space on the inside of the opening and closing parentheses. You can see this in the following snippet:

```
func GetString( id: Int )
```

The spaces keep it consistent and easy to read.

Function calling spacing

Calling a function also follows the exact same rules as defining a function. You can see this in the following code snippet:

```
Adder( num1: 10, num2: 5 )
```

It's easier to read this way, and aesthetically pleasing.

Square bracket spacing

When using square brackets, do not use any spaces on the inside of the opening and closing brackets. You can see this in the following code snippet:

```
let vectors : [[Int]] = [[1, 2, 3], [4, 5, 6]]
```

Although this is different from parentheses, it looks better this way. Line lengths should generally be no longer than 80 characters, but if it helps with the readability, then exceptions can be made.

Typecasting

When typecasting, always enclose the type in parentheses and not the variable, and use a single space inside the parentheses but not outside the closing parenthesis. You can see this in the following snippet:

```
( Int )age
```

This way, it's easier to read and aesthetically pleasing. It also makes figuring out the association between variables and casts easier when nested.

Naming conventions

This section will cover the naming conventions used throughout this book. Always use meaningful but not long names. Let the code be self-documenting as much as possible but **NEVER** at the cost of readability and maintenance (don't use really long variable names such as `scoreForThePlayerForLevel1`; instead, use something such as `scorePlayerLevel1`).

- **Classes/functions/methods** – PascalCase, for example, `EpicFunction`

- **Variables** – camelCase, for example, `epicVariable`
- **Constants** – ALL LETTERS UPPERCASE

All variables belonging to an object such as a class should start with an underscore – for example, `_localVariable`.

This is helpful because it keeps the code consistent, which improves efficiency, readability, and maintenance. It helps others understand your code better. It also looks nice.

Yoda conditions

When comparing variables and values, always compare the value to the variable and not the variable to the value. You can see this in the following code block:

```
if ( 10 == counter )
```

This is helpful because it prevents you from accidentally assigning a value to the variable instead of comparing it and thus results in a true value and makes the if statement equate to true.

Comments

This is an area that most people hate, and a lot of people miss out or leave until the end and then come to the realization that it's a lot of effort at that stage, so I would recommend commenting in the code as you go along.

When commenting in code, use // for single-line comments and /*   */ for multiline comments (the way your comment may vary based upon the programming language you are using) and all letters should be lowercase. You can see this in the following example:

```
/*
    This checks if the counter is 10
    If successful then inform user
*/
if ( 10 == counter )
{
    print( "Counter is 10" )
    // Print congratulation message
    print( "Well done" )
}
```

This is useful when trying to understand what your code does, especially if it is complex and/or you are reading it after a prolonged period of time. It is also useful when other people/programmers are trying to understand your code, as you may code differently to other people and comments can be crucial to helping them understand what is going on.

No magic numbers

A magic number is a number that looks like it is randomly placed in the code and doesn't have any context or obvious meaning. This is what we would call an anti-pattern since reading and understanding code becomes very difficult to maintain. It is important that the code is intentional and that, just at a quick glance, you are able to understand it; this is fundamental to code quality.

Use constants and variables. MAX_COUNT is used instead of an arbitrary number such as 10. You can see this in the following code snippet:

```
if ( MAX_COUNT == counter )
{
    print( "Counter is 10" )
    print( "Well done" )
}
```

This makes the code base easier to manage and understand.

In this section, we looked at the coding standards I use. I felt it very important to explain them not necessarily to persuade you to use the ones that I opt for, but so you have a reference point if you see anything in the book that seems different from your own normal standards. Above all, else be consistent with your standards and consistent with any teammates that are collaborating with you. This is the most important thing; the specific standards used are secondary, but the consistency is primary.

What is Swift?

In this section, we will cover what Swift is, its history, and how it works on a macro level. If you are an expert and just want to read about SwiftUI, feel free to skip this section.

Swift is a programming language created by Apple and the open source community. It is a general-purpose, compiled, and multi-paradigm programming language. It was released in 2014 as a replacement for Apple's previous language, Objective-C. Due to the fact that Objective-C had remained virtually the same since the early 1980s, it was missing many features that modern languages have. Hence, the creation of Swift began; it has taken the Apple developer ecosystem by storm and is a hugely popular programming language. It is demanded by companies all over the world with excellent remuneration offered to those that know how to leverage its immense features. According to the PYPL index seen in the following figure, Swift is in the top 10 most popular languages, making it a must-have in any programmer's arsenal:

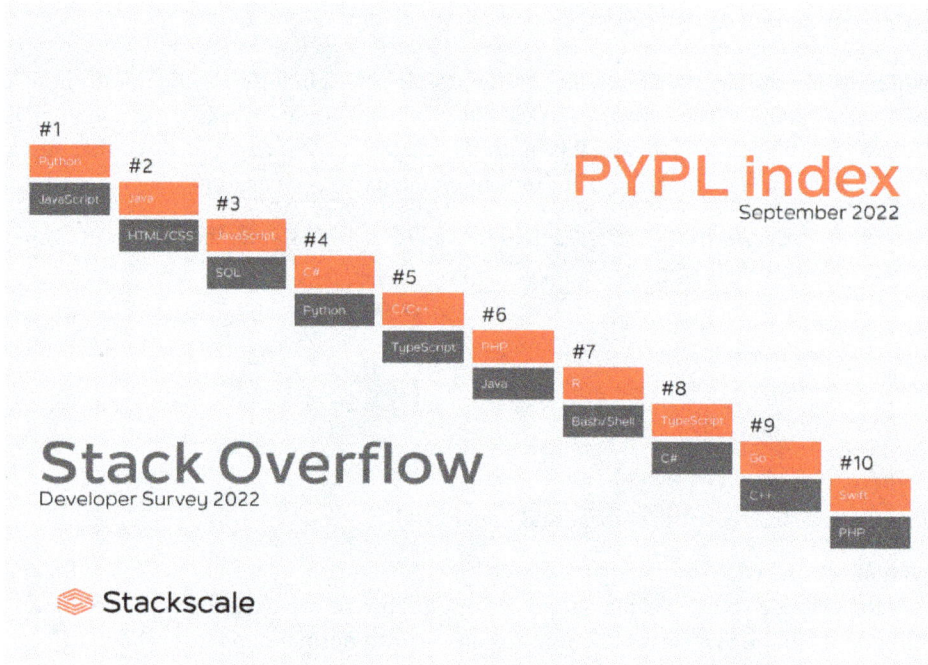

Figure 1.1 – Popularity of Programming Language index (Source: `https://www.stackscale.com/wp-content/uploads/2022/09/PYPL-index-popular-programming-languages-2022.jpg`)

Swift has been used to create many apps, including but not limited to the following:

- LinkedIn
- Firefox
- WordPress
- Wikipedia
- Lyft

Apple's Cocoa and Cocoa Touch frameworks work with Swift out of the box. Furthermore, it works flawlessly with Apple's previous programming language Objective-C, which has been used by developers for decades. This makes it one of the most modern but accessible languages around. Gone are the days when you would need to wait for frameworks to be released/updated for your chosen language; instead, thousands of projects already exist in Objective-C to use in the meantime.

Swift was built using the **Low-Level Virtual Machine** (LLVM) compiler framework, which is open source, is bundled with Xcode (since version 6), and was also released in 2014. It uses the Objective-C runtime library on Apple devices, thus allowing code written in C, Objective-C, C++, and Swift to work together in a single application. The following figure explains the relationship between programming languages such as Swift, LLVM, and the different architectures from a macro level:

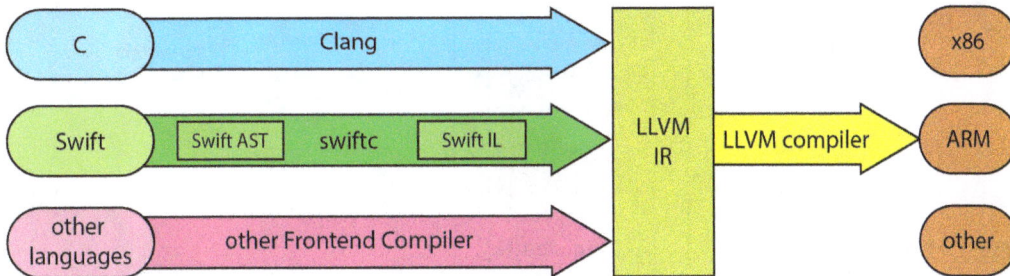

Figure 1.2 – LLVM and language relationship

(Source: `https://miro.medium.com/`
`max/1024/1*VWogVHhCagxopvAKVFjBeA.jpeg`)

LLVM uses Clang on the frontend, which is a compiler for programming languages such as C, C++, CUDA, or swiftc for Swift. This then turns the code into a format that LLVM uses to convert into the machine code, which is then run/executed on the hardware.

For more information on LLVM, feel free to use the following links:

- `https://llvm.org/`
- `https://en.wikipedia.org/wiki/LLVM`
- `https://github.com/llvm/llvm-project`
- `https://www.youtube.com/watch?v=BT2Cv-Tjq7Q&ab_channel=Fireship`
- `https://www.youtube.com/watch?v=IR_L1xf4PrU&ab_channel=tanmaybakshi`
- `https://www.youtube.com/watch?v=ZQds2aGHwDA&ab_channel=LexFridman`
- `https://www.youtube.com/watch?v=yCd3CzGSte8&ab_channel=LexFridman`

To understand the features available to you as a developer in Swift from a macro perspective, take a look at the following diagram:

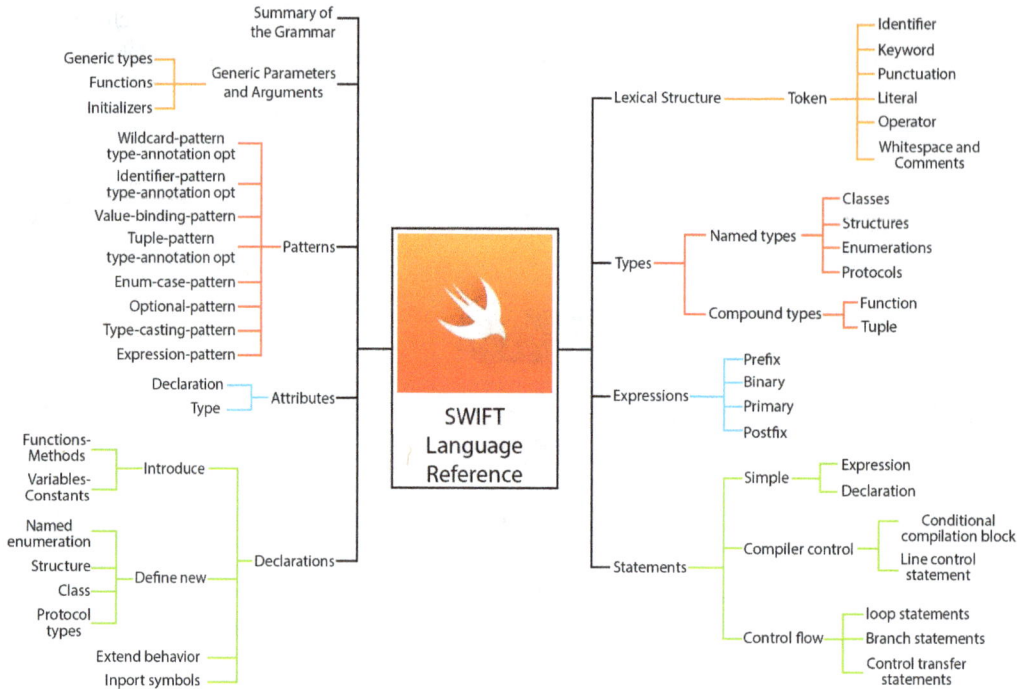

Figure 1.3 – Swift language reference

(Source: `https://gogeometry.com/software/swift/`
`swift-language-reference-mind-map.jpg`)

The preceding diagram is a mind map showing all the high-level features provided by Swift and their subfeatures and how they all link together.

In this section, we covered what Swift is, how it works, and its popularity.

What is SwiftUI?

In this section, we will cover what SwiftUI is and the features provided that we will leverage throughout this book to create our projects. If you feel comfortable with SwiftUI and just want to see projects, then feel free to skip the remainder of this chapter.

SwiftUI is a user interface framework built on top of the Swift programming language. It provides many components for creating your app's user interface; the following is a macro list of these components:

- Views and controls
- Shapes

- Layout containers
- Collection containers
- Presentation containers

In addition to the components in the preceding list, SwiftUI provides us with event handlers, allowing our apps to react to taps, gestures, and all other types of input they may receive from the user. The framework provides tools to manage the flow of data from the models to the views and controls that the end user interacts with.

Now we will look at the different core features of Swift, including examples that you can take away, modify, and use in your own projects.

Views and Controls

Views and controls are the foundational blocks of your application's UI. Using views, we can build the UI you want for your app. Its complexity can be whatever you desire, simple or immensely complex – it's totally up to you – and we will see this in more detail in the upcoming sections.

Views can be any of the following:

- Text
- Images
- Shapes
- Custom drawings
- A combination of all of these

Controls enable user interaction with APIs that adapt to the platform and context they are used in.

Shapes

Shapes in SwiftUI are 2D objects such as circles and rectangles. Custom paths can also be leveraged to set the parameters of your own shape/structure; we will see shapes in more detail in the coming sections.

Shapes provide features to add styling, including but not limited to the following:

- Environment-aware color
- Rich gradients
- Material effects in the foreground
- Background
- Outlines for your shapes

Layout Containers

Layouts have the job of organizing the elements of your app's UI. Stacks and grids are used to update and modify the positions of the child views that are within them in response to changes in content or interface dimensions. Layouts can be nested within one another; this can be done to as many levels as desired, thus allowing you to create complex layouts. Custom layouts can also be designed for further flexibility; we will take a look at layout containers in more depth later on.

Collection Containers

Collections can be used to assemble dynamic views with complex functionality. For example, you can create a List view that allows you to scroll through a large set of data. The list automatically provides basic functionality, but in addition to this, you can add other functionality with minimal configuration, such as swiping, double tapping, and pull-to-refresh.

If only a simple grid or stack configuration is required, use a Layout container instead; we will look at code examples to further illustrate these.

Presentation Containers

Presentation containers are used to provide structure to your app's UI. This provides users with easier navigation for jumping around the app. This is extremely useful as the complexity of it increases and it contains more views. For example, you can enable navigating backward and forward through a set of views using a `NavigationStack`, and choosing which view to display from a tab bar using a `TabView`.

In this section, we covered the different features provided by SwiftUI on a macro level; we also discussed what sub-features they have and their uses.

Understanding and implementing views

In this section, we will look at views and how we can implement them in SwiftUI. We will also take a look at combining these views.

Views are the fundamental building blocks of an application's user interface. A view object renders content within its rectangle bounds and handles any interactions with that content.

In the following sections, we will show the source code and examples for each type of view. If you would like further information, visit Apple's documentation at `https://developer.apple.com/documentation/uikit/views_and_controls`.

What are text views?

It is very common to need to display text in our app, and we do this by using a text view, which draws a string. By default, it has a font assigned to it that is best for the platform it is being displayed on; however, you can change the font using the font (_:) view modifier.

The following snippet shows the code used to implement a text view:

```
var body: some View
{
    VStack
    {
        Text( "Hello World" )
    }
    .padding( )
}
```

The text view was inserted into a `VStack` for padding purposes, but this is not a requirement.

The preceding code shows how to simply display a string using the Text view by passing a string of what you need to display.

The following figure shows the output of the preceding code:

Figure 1.4 – Text view preview

The next section will cover image views and how to implement them.

What are Image views?

Image views can be used to render images inside your SwiftUI layouts. Images are an excellent way of providing more context and improving the overall user experience. Image views can load images from your bundle, from system icons, from a `UIImage`, and more, but loading from your bundle and system icons will be the most commonly used method.

The following snippet shows the code used to implement an image view:

```
var body: some View
{
```

```
    VStack
    {
        Image( systemName: "cloud.heavyrain.fill" )
    }
    .padding( )
}
```

In the preceding code, we implement an Image View with a system icon but you can easily specify your own image file. We will do this later in this chapter.

In this example, a system icon was used but the process is similar for using your bundle and a UIImage.

The following figure shows the output of the preceding code:

Figure 1.5 – Image view preview

The next section will cover the different shape views available to us and the code to use them.

What are shape views?

SwiftUI provides us with five primitive shapes that are commonly used. These shapes are rectangles, rounded rectangles, circles, ellipses, and capsules. The last three are very subtly different in how they behave based on what sizes you provide.

The following code shows how you can simply implement any of the following shapes:

- Rectangle
- RoundedRectangle
- Capsule
- Ellipse
- Circle

Here's the snippet:

```
var body: some View
{
    VStack
    {
        Rectangle( )
            .fill( .white )
            .frame( width: 128, height: 128 )

        RoundedRectangle( cornerRadius: 30, style: .continuous )
            .fill( .blue )
            .frame( width: 128, height: 128 )

        Capsule( )
            .fill( .red )
            .frame( width: 128, height: 50 )

        Ellipse( )
            .fill( .orange )
            .frame( width: 128, height: 50 )

        Circle( )
            .fill( .yellow )
            .frame( width: 128, height: 50 )
    }
    .padding( )
}
```

The shapes were inserted into a VStack to arrange vertically and have padding around them.

The preceding code shows how to implement the different shape views and the parameters needed for each one.

The following figure shows the output of the preceding code:

Figure 1.6 – Shape views

In the next section, we will take a look at how we can use the views we have covered to combine and create more complex views.

What are custom and combination views?

For all frontend developers, one of the most crucial aspects of the development process is the implementation of the UI. We can create a simple UI with a combination of pre-made in-built views that we have leveraged thus far, but this is sometimes not enough; there are often cases when developers need to draw custom views programmatically to meet UI requirements, and if we are not able to draw those, it creates a problem. We are able to leverage the power of SwiftUI to create custom views, which are effectively a combination of the other views we learned about previously.

The following code shows us an implementation of multiple views to create our own custom view that displays a mini profile:

```
import SwiftUI

struct Employee
{
```

```
    var name: String
    var jobTitle: String
    var emailAddress: String
    var profilePicture: String
}

struct ProfilePicture: View
{
    var imageName: String

    var body: some View
    {
        Image( imageName )
            .resizable( )
            .frame( width: 100, height: 100 )
            .clipShape( Circle( ) )
    }
}

struct EmailAddress: View
{
    var address: String

    var body: some View
    {
        HStack
        {
            Image( systemName: "envelope" )
            Text( address )
        }
    }
}

struct EmployeeDetails: View
{
    var employee: Employee

    var body: some View
    {
        VStack( alignment: .leading )
        {
            Text( employee.name )
                .font( .largeTitle )
```

```
                    .foregroundColor( .primary )
                Text( employee.jobTitle )
                    .foregroundColor( .secondary )
                EmailAddress( address: employee.emailAddress )
            }
        }
}

struct EmployeeView: View
{
    var employee: Employee

    var body: some View
    {
        HStack
        {
            ProfilePicture( imageName: employee.profilePicture )
            EmployeeDetails( employee: employee )
        }
    }
}

struct ContentView: View
{
    let employee = Employee( name: "Frahaan Hussain", jobTitle: "CEO
& Founder", emailAddress: "frahaan@hussain.com", profilePicture:
"FrahaanHussainIMG" )

    var body: some View
    {
        EmployeeView( employee: employee )
    }
}

struct ContentView_Previews: PreviewProvider
{
    static var previews: some View
    {
        ContentView( )
    }
}
```

The preceding code shows how we can combine views to create custom views and how we can combine custom views to create more complex views. This also makes the mini views reusable.

The following figure shows the output of the preceding code:

Figure 1.7 – Custom view (mini profile)

In this section, we looked at combining the views from the previous sections to make reusable and combinable objects and create more complex views.

In the next section, we will take a look at layouts to help organize the content in our app.

Understanding and implementing layouts

This section will cover how we can arrange our views using layouts for a more dynamic user experience.

SwiftUI layouts allow us as developers to arrange views in your app's interface using the layout tools provided. Layouts tell SwiftUI how to place a set of views, and how much space it needs to do so to provide the desired layout.

Layouts can be but are not limited to any of the following:

- Lazy stacks
- Spacers:
  - `ScrollViewReader`
- Grids:
  - `PinnedScrollableViews`

In the following sections, we will show you the source code and examples for each type of layout.

> **Note**
> If you would like further information, visit Apple's documentation: `https://developer.apple.com/documentation/uikit/view_layout`.

What are lazy stacks?

Lazy stacks are views that arrange their children in a line that expands vertically, creating items only as needed.

SwiftUI provides two different types of lazy stacks, LazyVStack and LazyHStack. By default, VStack and HStack load all the content upfront, which will be slow if you use them inside a scroll view, as these views can contain a lot of content. If you want to load content in a lazy fashion, so it is only loaded when it appears in the view, and not when the view is generally visible but the content is not, you should use LazyVStack and LazyHStack as appropriate.

The word lazy, as explained by Apple, refers to the stack view not creating items until they are needed. What this means to you is that the performance of these stack views is already optimized by default.

LazyVStack and LazyHStack are only available in iOS 14.0+, iPadOS 14.0+, macOS 11.0+, Mac Catalyst 14.0+, tvOS 14.0+, and watchOS 7.0+. More information can be found at the following links:

- LazyVStack – https://developer.apple.com/documentation/swiftui/lazyvstack
- LazyHStack – https://developer.apple.com/documentation/swiftui/lazyhstack

The following code shows how we can use a LazyVStack to organize views vertically while also being efficient with large sums of data:

```
import SwiftUI

struct ContentView: View
{
    var body: some View
    {
        ScrollView
        {
            LazyVStack
            {
                ForEach( 1...1000, id: \.self )
                {
                    value in
                    Text( "Line \( value )" )
                }
            }
        }
        .frame( height: 256 )
    }
}
```

```
struct ContentView_Previews: PreviewProvider
{
    static var previews: some View
    {
        ContentView( )
    }
}
```

The preceding code implements a `LazyVStack` with `1000` text views. The text views are added using a loop to make it easier and more efficient.

The following figure shows the output of the preceding code:

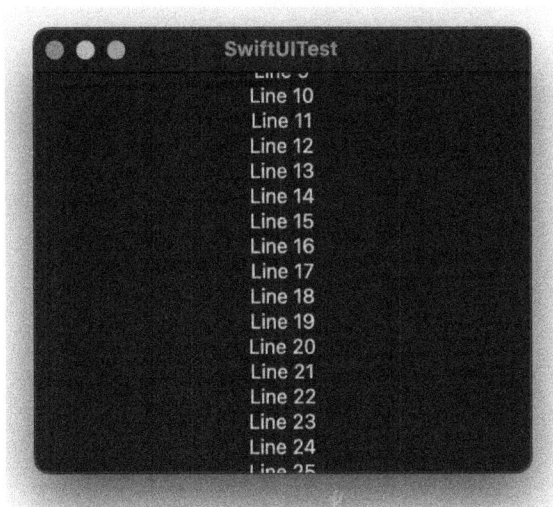

Figure 1.8 – Lazy stacks

In the next section, we will take a look at spacers, which help us space out our content.

What are spacers?

A spacer creates a view that is adaptive with no content that expands as much as it can. For example, when placed within an `HStack`, a spacer expands horizontally as is allowed by the stack, moving views out of the way, within the size limits of the stack.

In the following code, we implement text and spacers:

```
import SwiftUI

struct ContentView: View
```

```
{
    var body: some View
    {
        Text( "Label 1" )

        Spacer( ).frame( height: 64 )

        Text( "Label 2" )
    }
}

struct ContentView_Previews: PreviewProvider
{
    static var previews: some View
    {
        ContentView( )
    }
}
```

The preceding code used a spacer with a height of 64 to separate the two text views.

The following figure shows the output of the preceding code:

Figure 1.9 – Spacer preview

In the next section, we will look at ScrollViewReader, which enables us to move to any location.

What are ScrollView and ScrollViewReader?

ScrollView allows users to view content within a scrollable region. The user can perform platform-specific scroll gestures to adjust the visible portion of the content. ScrollView can scroll both horizontally and vertically but does not support zooming. If you want to programmatically move `ScrollView` to a specific location, you should add a `ScrollViewReader` inside it. This provides a method called `scrollTo()`, which moves to any view inside the parent `ScrollView`, simply by providing its anchor.

All of this can be achieved with a few simple lines of code. You can see this code as follows:

```swift
import SwiftUI

struct ContentView: View
{
    var body: some View
    {
        let colors: [Color] = [.red, .green, .blue, .white, .yellow]

        ScrollView
        {
            ScrollViewReader
            {
                value in
                Button( "Go to Number 45" )
                {
                    value.scrollTo( 45 )
                }
                .padding( )

                ForEach( 0..<1000 )
                {
                    i in
                    Text( "Example \( i )" )
                        .font( .title )
                        .frame( width: 256, height: 256 )
                        .background( colors[i % colors.count] )
                        .id( i )
                }
            }
        }
        .frame( height: 512 )
    }
}

struct ContentView_Previews: PreviewProvider
{
    static var previews: some View
```

```
    {
        ContentView( )
    }
}
```

The preceding code implemented `ScrollViewReader` inside `ScrollView`. Inside that, we added a button to scroll to the text view, which has an `id` value of `45`. We used a loop to display `1000` text views with unique IDs.

The following figure shows the output of the preceding code:

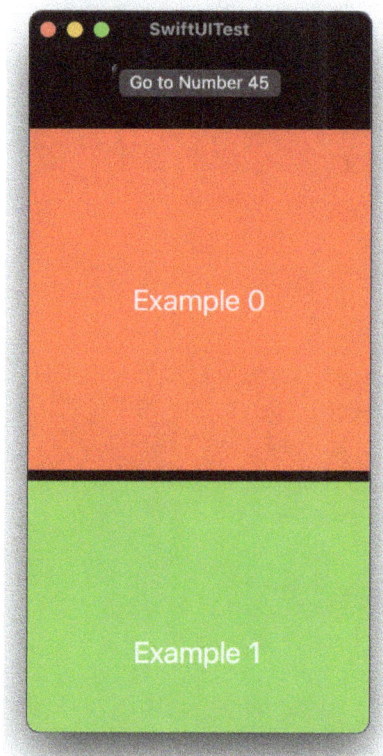

Figure 1.10 – ScrollViewReader preview

In the next section, we will look at grids, which allow us to organize our content in a table-like fashion.

What are grids?

When SwiftUI was first released, it didn't come with a collection view built in. Developers were left with one of two options – either to build their own or use a third-party solution. In WWDC 2020, Apple introduced new features for the SwiftUI framework. One of them was to address the need for grid views. SwiftUI now provides two new components, `LazyVGrid` and `LazyHGrid`. One is for vertical grids and the other is for horizontal grids.

The following code shows how we can implement a grid with two rows:

```swift
import SwiftUI

struct ContentView: View
{
    let items = 1...50

    let rows =
    [
        GridItem( .fixed( 32 ) ),
        GridItem( .fixed( 32 ) )
    ]

    var body: some View
    {
        ScrollView( .horizontal )
        {
            LazyHGrid( rows: rows, alignment: .center )
            {
                ForEach( items, id: \.self )
                {
                    item in
                    Image(systemName: "\( item ).circle.fill" )
                        .font( .largeTitle )
                }
            }
            .frame( height: 128 )
        }
    }
}

struct ContentView_Previews: PreviewProvider
{
    static var previews: some View
    {
        ContentView( )
    }
}
```

In the preceding code snippet, we implement a grid of images set over two rows using a loop. `GridItem` components were used; Apple says these are descriptions of rows or columns in a lazy grid. What does this actually mean? Well, it's essentially a method for specifying how many columns/rows we want, thus setting the layout pattern. `LazyGrid` uses this layout pattern when iterating through the items that are displayed and positioning them accordingly. If you are coming from a web background, you can think of it like the grid systems in responsive websites.

The following figure shows the output of the preceding code:

Figure 1.11 – Grid preview

The next section will cover `PinnedScrollableViews`, which allow certain views to stick on the page as others scroll past them.

What is PinnedScrollableView?

SwiftUI can provide a `PinnedScrollableView` inside a `ScrollView`. Pinned views are sticky views and can be applied to either a header or a footer.

The following code shows how we can pin a view to provide context to other views as they scroll past it:

```
import SwiftUI

struct MyCell: View
{
    var body: some View
    {
        VStack
        {
            Rectangle( )
                .fill( Color.red )
                .frame( width: 128, height: 128 )
            HStack
```

```
                {
                    Text( "Line text" )
                        .foregroundColor( .yellow )
                        .font( .headline )
                }
                Text( "PinnedScrollableViews" )
                    .foregroundColor( .green )
                    .font( .subheadline )
            }
        }
}

struct ContentView: View
{
    var stickyHeaderView: some View
    {
        RoundedRectangle( cornerRadius: 25.0, style: .continuous )
            .fill( Color.gray )
            .frame( maxWidth: .infinity )
            .frame( height: 64 )
            .overlay(
                Text( "Section" )
                    .foregroundColor( Color.white )
                    .font( .largeTitle )
            )
    }
    var body: some View
    {
        NavigationView
        {
            ScrollView
            {
                LazyVStack( alignment: .center, spacing: 50,
pinnedViews: [.sectionHeaders], content:
                {
                    ForEach( 0...50, id: \.self )
                    {
                        count in
                        Section( header: stickyHeaderView )
                        {
                            MyCell( )
                        }
                    }
                } )
```

```
                    }
                }
            }
        }

    struct ContentView_Previews: PreviewProvider
    {
        static var previews: some View
        {
            ContentView( )
        }
    }
```

In the preceding code snippet, we implemented a sticky view, which acted as the section header as the other views dynamically moved past it, but the sticky view moved off the screen when another section was reached, and it had its own sticky view.

The following figure shows the output of the preceding code:

Figure 1.12 – PinnedScrollableViews preview

In this section, we covered the different layouts we have access to. These allow us to organize the data in a more pleasing fashion. In the next section, we will look at the different device previews for the projects that we will create throughout this book.

Device Previews

One of the many hurdles we will overcome is the differences between the four main Apple product categories. These categories are as follows:

- Mac – iMac, Mac Pro, MacBook, anything that runs macOS, even Hackintosh
- iPad – Mini, regular, Air, Pro, all of them
- iPhone – Mini, Pro, Pro Max, all iPhones
- Apple Watch – small, big, old, or new

The obvious thing you will notice is that they get smaller and smaller. They naturally have different purposes; an Apple Watch won't replace a Mac and vice versa. That is why the next eight chapters are grouped into pairs, one for each product category. We will uncover the design decisions and restrictions we have when creating our applications. Let's take a look at the settings in each product category:

Mac

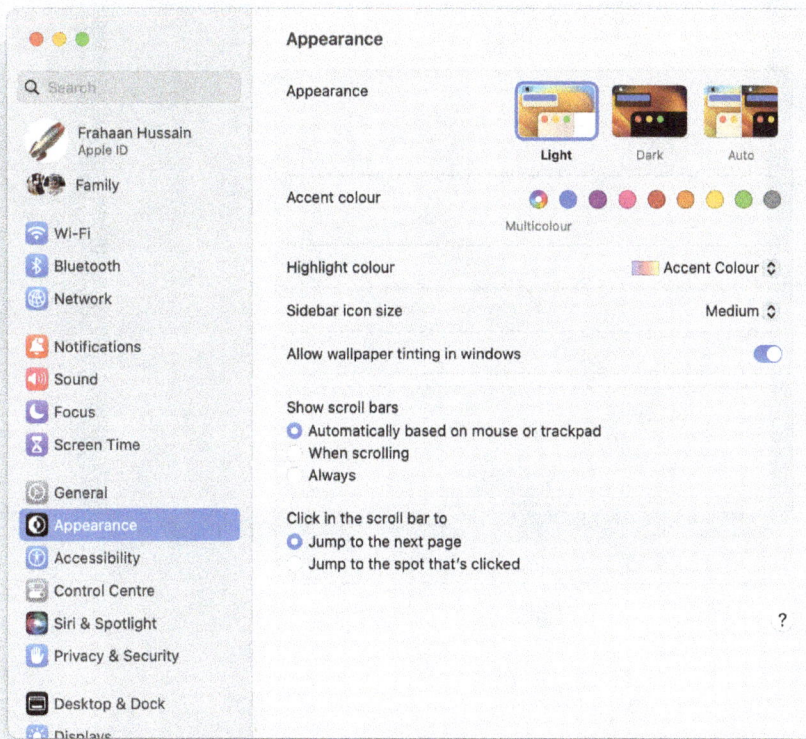

Figure 1.13 – Mac settings

iPad

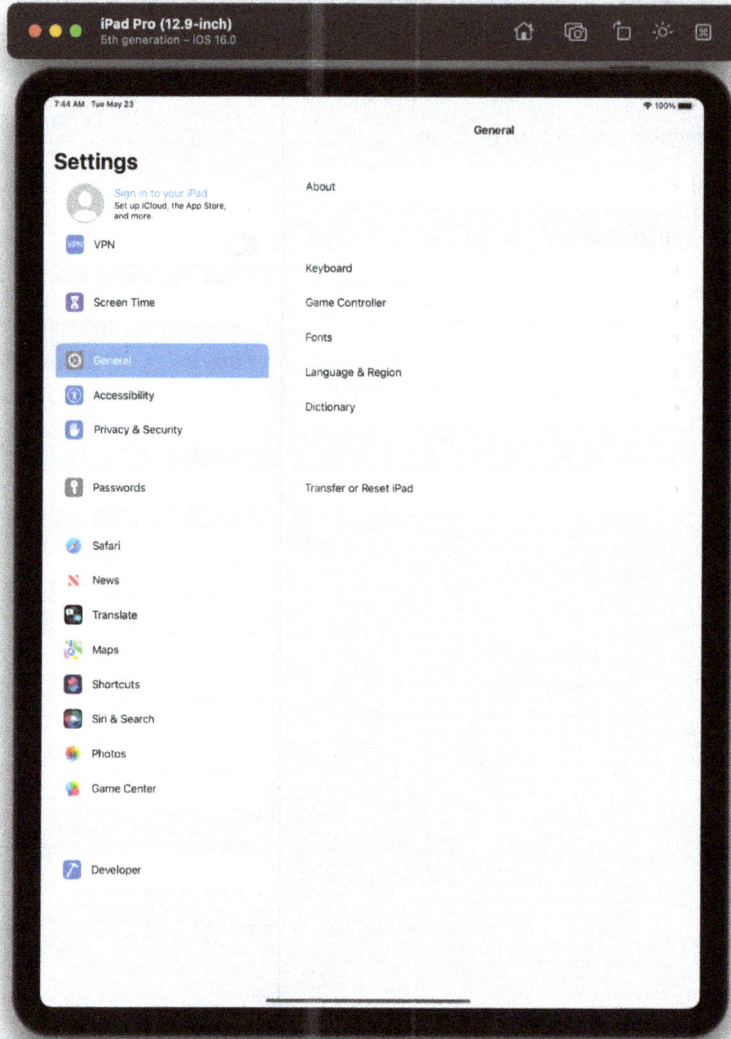

Figure 1.14 – iPad settings

iPhone

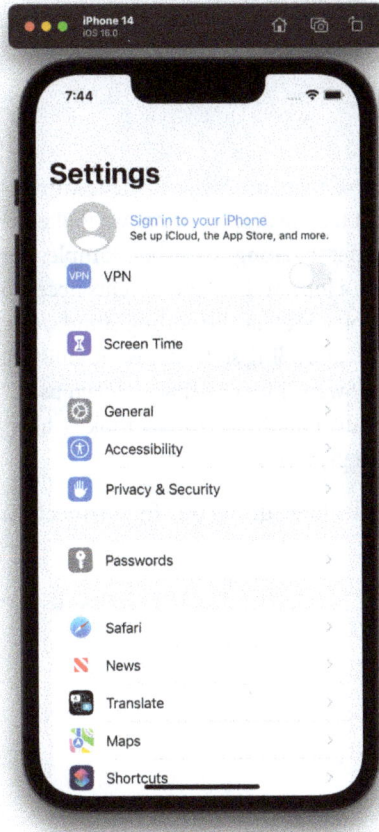

Figure 1.15 – iPhone settings

Apple Watch:

Figure 1.16 – Apple Watch settings

At a quick glance, it is immediately apparent that there are differences. These are the very differences we will discuss during the remainder of this book. The next section will summarize this chapter before heading over to our first project.

Summary

In this chapter, we covered the history of Swift and SwiftUI, the features provided to us on a macro level, and how Swift and SwiftUI work on a technical level. Then, we looked at the difference between Swift and SwiftUI and what features are offered, along with code samples for you to take away and use in your own projects. After that, we took a look at the requirements necessary for developing applications with Swift and SwiftUI. Then, we took a look at the coding standards that are used throughout this book, providing a reference point for any coding style that is unfamiliar to your own. Then, we looked at the views and controls provided by SwiftUI for creating our own user experiences, including custom views by combining the fundamentals. Finally, we took a look at how we can organize these views using layouts and checked the device previews too.

In our next chapter, we'll take a look at designing our first project, the tax calculator app that we will create.

2

iPhone Project – Tax Calculator Design

In the previous chapter, we did a recap of Swift and SwiftUI. We looked at the requirements, the coding standards used, and the basics of SwiftUI components. We will use these in the following chapters.

In this chapter, we will work on the design of our first project, a tax calculator. We will assess the requirements for designing such an application and discuss the design specifications, allowing us to get a better understanding of what is required and how it will all fit together. Then, we will start our application's coding process to build out the UI, which will be connected together allowing the application to fully function in the next chapter. This project will teach us the foundations of SwiftUI components and how to interact with external code bases. We will discuss all of this in the following sections:

- Technical Requirements
- Understanding the Design Specifications
- Building the Calculator UI

By the end of this chapter, you will have a better understanding of what is required and the design of our application. You will also have a skeleton UI that will be used as the foundation for making the calculator work in the next chapter.

In the next section, we will provide clarity on the specifications of our application's design and look at mockups of what the app will look like.

Technical requirements

This chapter requires you to download Xcode version 14 or above from Apple's App Store.

To install Xcode, just search for Xcode in the App Store, then select and download the latest version. Open Xcode and follow any additional installation instructions. Once Xcode has opened and launched, you're ready to go.

Version 14 of Xcode has the following features/requirements:

- Includes SDKs for iOS 16, iPadOS 16, macOS 12.3, tvOS 16, and watchOS 9.

- Supports on-device debugging in iOS 11 or later, tvOS 11 or later, and watchOS 4 or later.

- Requires a Mac running macOS Monterey 12.5 or later.

Download the sample code from the following GitHub link:

```
https://github.com/PacktPublishing/Elevate-SwiftUI-Skills-by-Building-
Projects
```

In the next section, we will provide clarity on the specifications of our application's design and look at mockups of what the app will look like.

Understanding the Design Specifications

In this section, we will look at the design specifications of our tax calculator application. This section describes the features we are going to implement in our tax calculator app. The best method for figuring out the features required is to put yourself in the user's shoes to determine how they will use the app and break it into individual steps.

The features of our app are as follows:

- Income entry – the ability to enter an income.

- Salary summary – a summary of how much is going to be taxed and how much is left as income.

- Tax breakdown – a breakdown of how much tax is paid on a given salary, that is, tax brackets.

- Different taxes – the ability to calculate a breakdown for different types of taxes, such as income, property flipping, inheritance, and stamp duty.

- Tax geography – the ability to calculate a breakdown for taxes in different geographies, including countries and states.

- Combination of the previous two – the ability to calculate a breakdown for different taxes in different geographies.

- User system – allows users to create an account to store tax calculations, see how the tax has changed over time relative to new tax laws, and so on.

Now that we have listed the ideal features we would like, next, it is important for us to determine which features are absolutely crucial. To do this, we must understand the end use of our product. For me, the purpose of creating this tax calculator is not to release it and have it serve millions of people but to be a personal project for our use. It is to demonstrate a basic implementation of SwiftUI within the context of this book. Based on that, I know that all the features are not required; actually, it would be useful if some were omitted and assigned as extra tasks for you as the developer to undertake. Based on all of this, the following are the core features we will be implementing:

- Income entry
- Salary summary
- Tax breakdown

The rest of the features will be left for you to implement as an exercise once you have completed this and the next chapter. The next section will cover the acceptance criteria for our application.

Acceptance criteria

We will discuss the mandatory requirements for our application that we absolutely want to see in the end product at the end of the next chapter. If possible, we should try to make them measurable. Let's do this right now:

- Error detection:

  - **Not a Number (NaN)** values
  - Values that are equal to or less than 0

- Provides the before- and after-tax salaries
- A pie chart to illustrate the breakdown visually
- Progress bars to further expand upon the breakdown
- Navigation to allow users to switch between the pages effortlessly

Develop test cases in which the application's acceptance criteria will be tested. Using this method allows you to see the conditions in which the application will be used by the end user and the level of functionality that needs to be attained for it to be considered successful.

Wireframe

One of the most useful tools for designing layouts is wireframing. A wireframe is an overview of how the layout will look. The following figure shows what the front page of our app would look like using a wireframe:

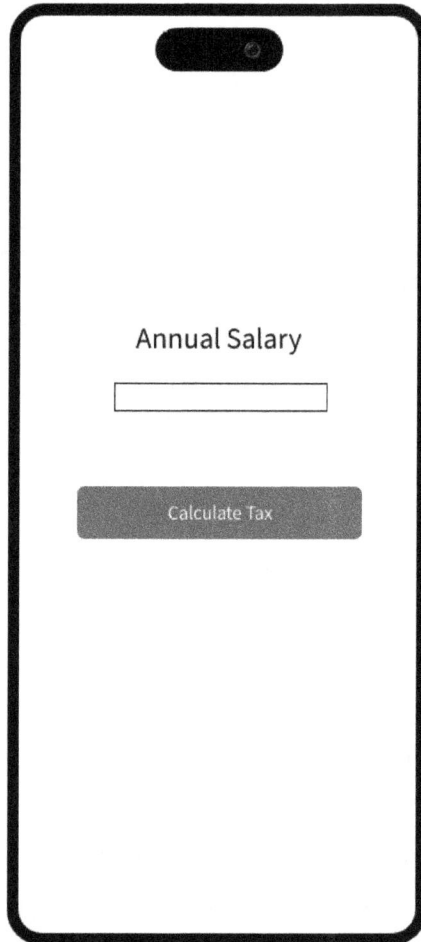

Figure 2.1 – Front page wireframe preview

The following figure shows the wireframe of what our results page would look like:

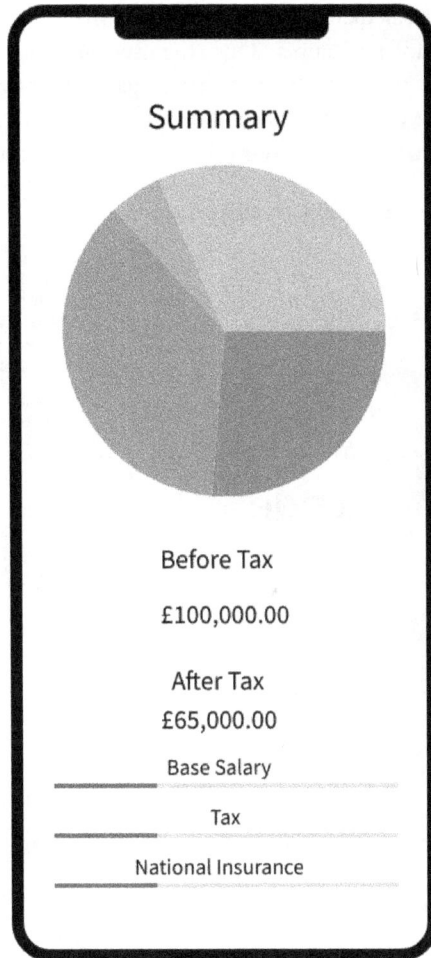

Figure 2.2 – Results page wireframe preview

In the next section, we will build the interface for our application and make sure it looks the way we designed it in the wireframes. Though we will build it the same way, there can be small differences. This will serve as the foundation for connecting it all together in the next chapter.

Building the calculator UI

We will now build the UI for the calculator app. There are two main parts to the calculator, the first being the front page, which is loaded on launch. Once the user inputs an income and hits **Calculate Tax**, they are taken to the results page, which is the second part. On this page, the results of the tax calculation and a breakdown of it will be displayed. Naturally, we will start off with the first part, the front page, but before even that, we will create our project. Follow these steps:

1. Open Xcode and select **Create a new Xcode project**:

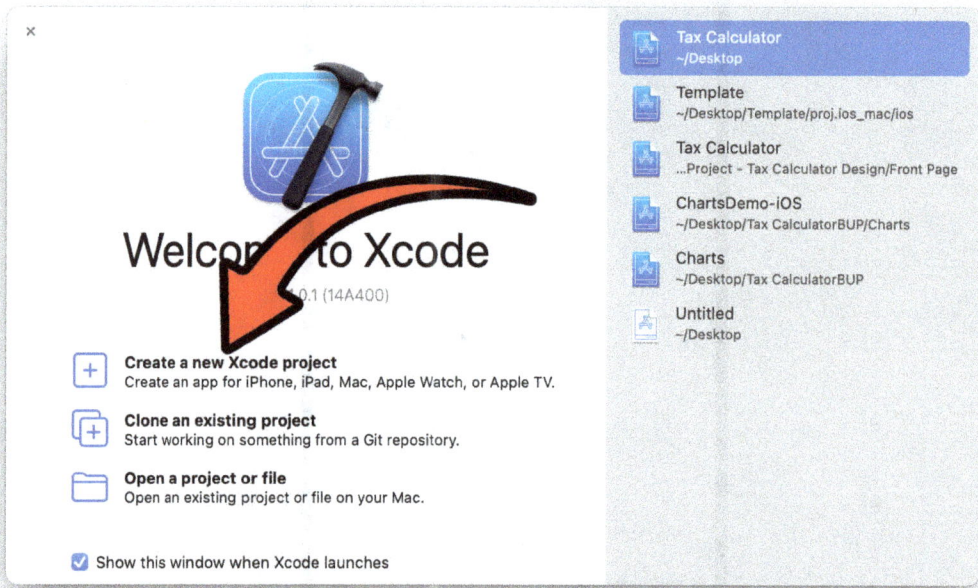

Figure 2.3 – Create a new Xcode project

2. Now, we will choose the template for our application. As we are creating an iPhone application, we will select **iOS** from the top and then select **App** and click **Next**:

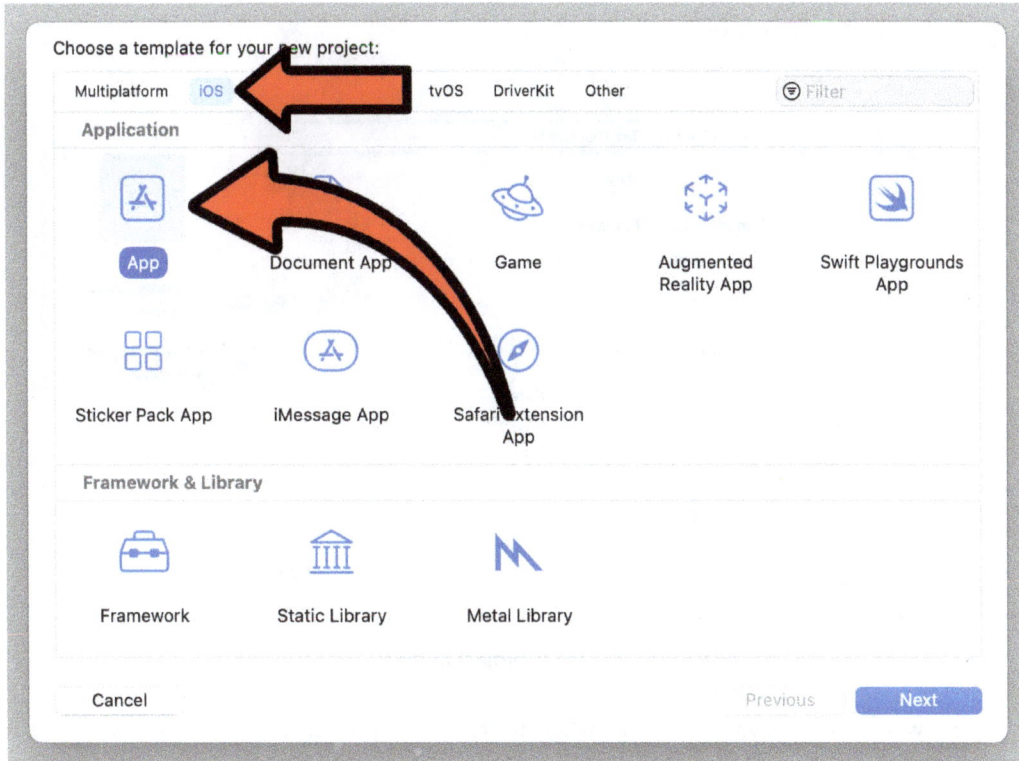

Figure 2.4 – Xcode project template selection

3. We will now choose the options for our project. Here, there are only two crucial things to select/ set. Make sure **Interface** is set to **SwiftUI**; this will be the UI our system will use. Set **Language** to **Swift**; this is the programming language used for our application:

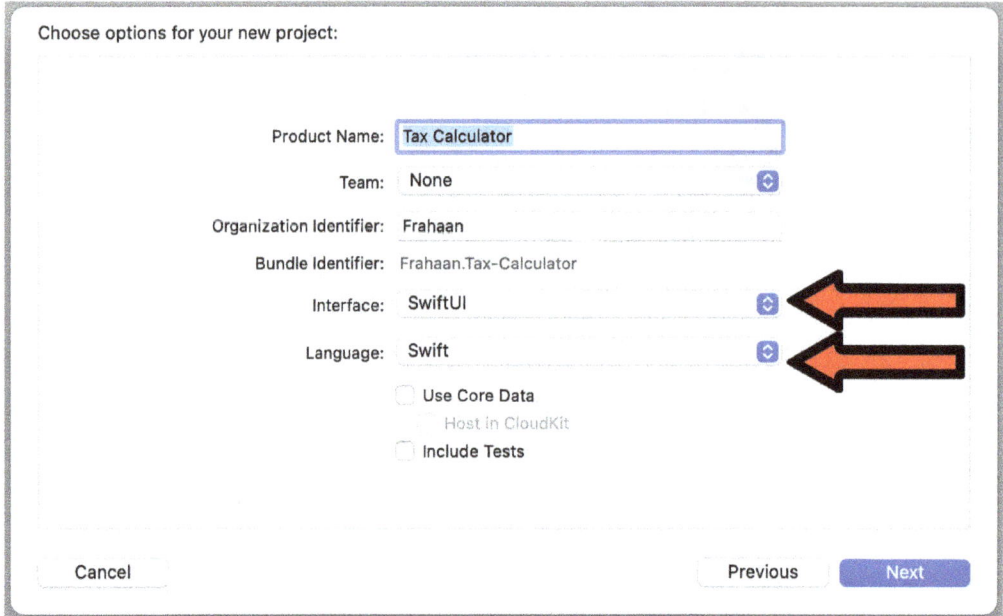

Figure 2.5 – Xcode project options

4. Once you press **Next**, you can choose where to create your project, as seen in the following screenshot:

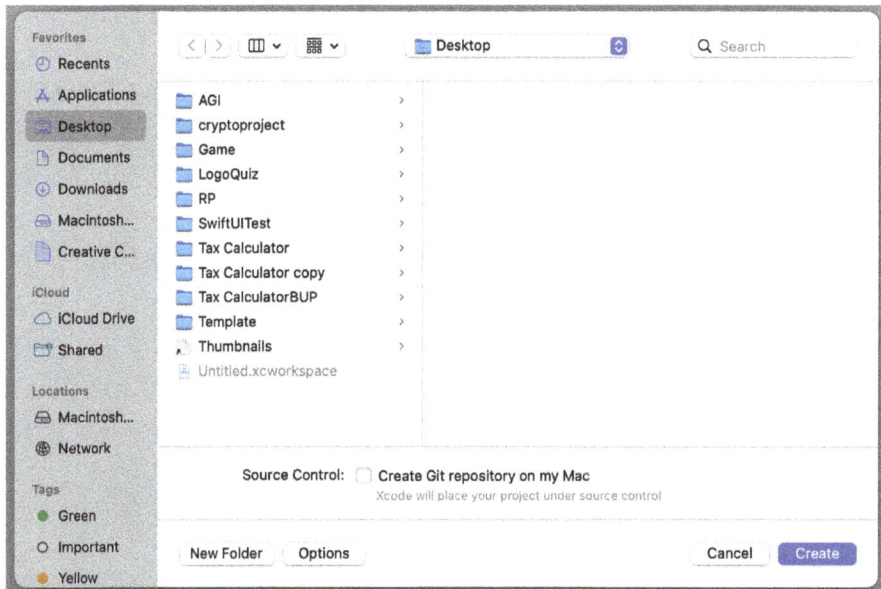

Figure 2.6 – Xcode project save directory

5. Once you have found the location you would like to create the project in, click on **Create** at the bottom right. Xcode shows your project in all its glory, as seen in the following screenshot:

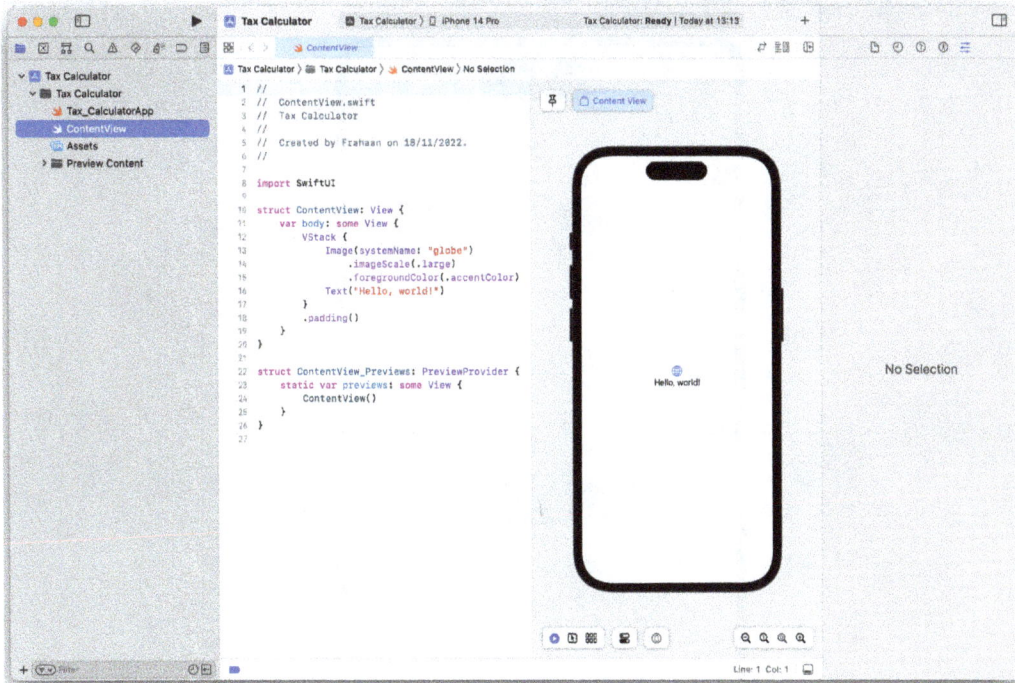

Figure 2.7 – New Xcode project overview

In the next section, we will implement the front page of our application using SwiftUI and further look at the Xcode IDE as we do so.

Front page

In this section, we will implement the front page's UI. As a reminder, here is what it will look like:

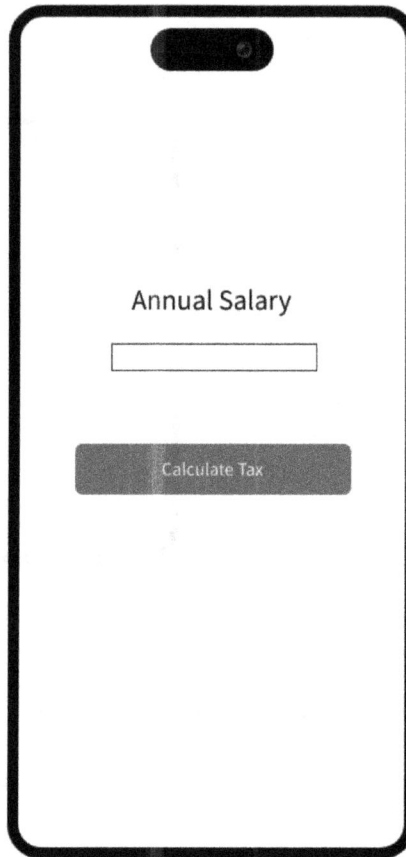

Figure 2.8 – Front page wireframe preview

There are three main elements on the front page. As a little task, see whether you can figure out what they are. Don't worry if you don't know the exact UI component names; we will look at these components in the following sections.

Text

A Text component is one of the simplest components offered by SwiftUI. It allows you to display a string of characters/numbers, which is very useful for headings and providing information. We will use it to provide context to the next component, which is TextField. Without the Text component, the user doesn't know the purpose of the TextField. The following figure shows the label on the front page, telling the user what the following text input field is used for:

Annual Salary

Figure 2.9 – Front page label

TextField

TextFields allow the user to input text that can consist of numbers and any character. Our TextField will be used to input a number, hence it will only accept numbers. This is a feature we will configure. Some applications put background text in the text field providing context to the purpose of the TextField; however, we have opted for a label component to provide context and do not require this. The following figure shows the TextField that the user can use to input their salary:

Figure 2.10 – Front page TextField

Button

Buttons are used when you want the user to explicitly trigger some functionality. In our case, we want the user to press the button when they are ready to calculate their tax calculation. Naturally, we as developers must error-check this to check whether the button can be pressed when the TextField is empty or if the wrong type of data has been inputted in the TextField. We will handle that with an error message instead of displaying the tax calculation. If you take a look at the following screenshot, you will see what this button looks like:

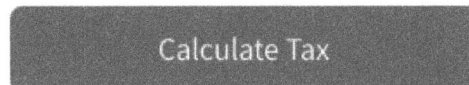

Calculate Tax

Figure 2.11 – Front page button

In the next section, we will add the elements we discussed previously using SwiftUI into our application.

Adding Front Page Components

In this section, we will add the components we listed previously to create our front page. Look for the ContentView file, which can be found in the **Project navigator**, usually on the left, as shown in the following screenshot:

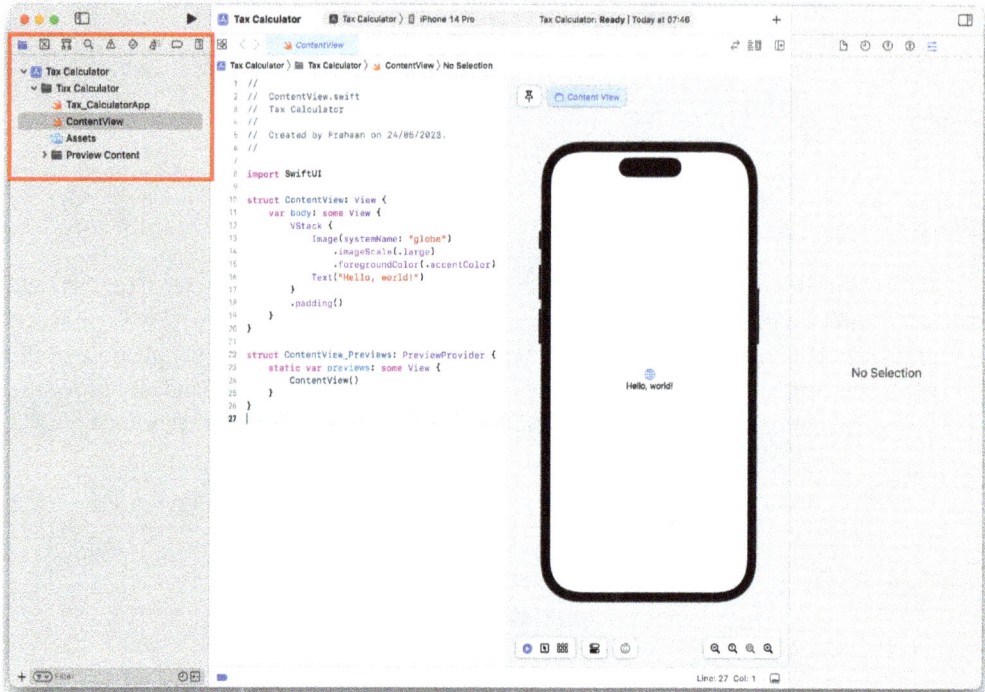

Figure 2.12 – Project navigator

See the following code and add it to the `ContentView` file:

```swift
import SwiftUI

struct ContentView: View
{
    @State private var salary: String = ""

    var body: some View
    {
        VStack
        {
            Text( "Annual Salary" )

            TextField( "Salary", text: $salary )

            Button
            { }
            label:
            { Text("Calculate Tax") }
        }
```

```
        .padding( )
    }
}

struct ContentView_Previews: PreviewProvider
{
    static var previews: some View
    {
        ContentView( )
    }
}
```

Using the preceding code, we are able to render a `Text`, `TextField`, and `Button`. This will form the basis of allowing the user to enter their salary and click the button to calculate the tax breakdown. We use a variable called `salary` to store TextField's data. Let's take a look at the end result:

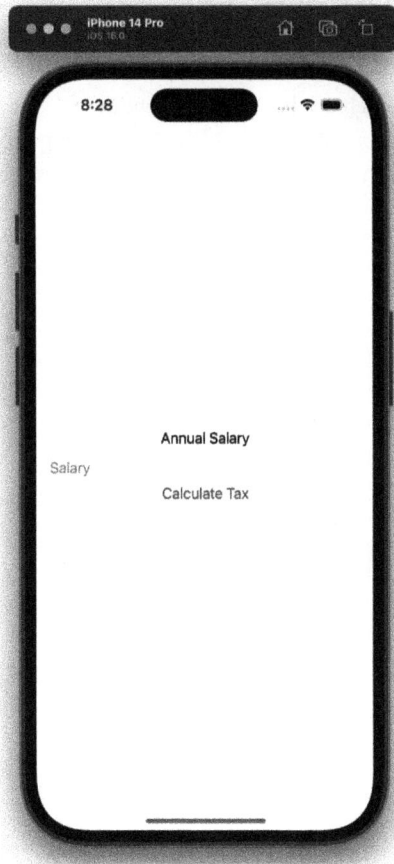

Figure 2.13 – Elements without styling preview

As you can see, the Text component looks pretty good, but the TextField has no obvious boundaries. I put a placeholder inside it as without it, the user wouldn't even know where the TextField is. Next, the Button has the wrong styling. Let's fix both of these with the following updated code:

```
VStack
{
    Text ( "Annual Salary" )

    TextField( "", text: $salary )
        .border ( Color.black, width: 1 )

    Button
    { }
    label:
    { Text ("Calculate Tax") }
        .buttonStyle ( .borderedProminent )
}
.padding ( )
```

Using the preceding code, we added a black border with a width of 1 to the TextField and removed the placeholder text. Next, we added a button style to the Text component of the button. We used the borderedProminent style, which is exactly what we need. All these changes result in the following:

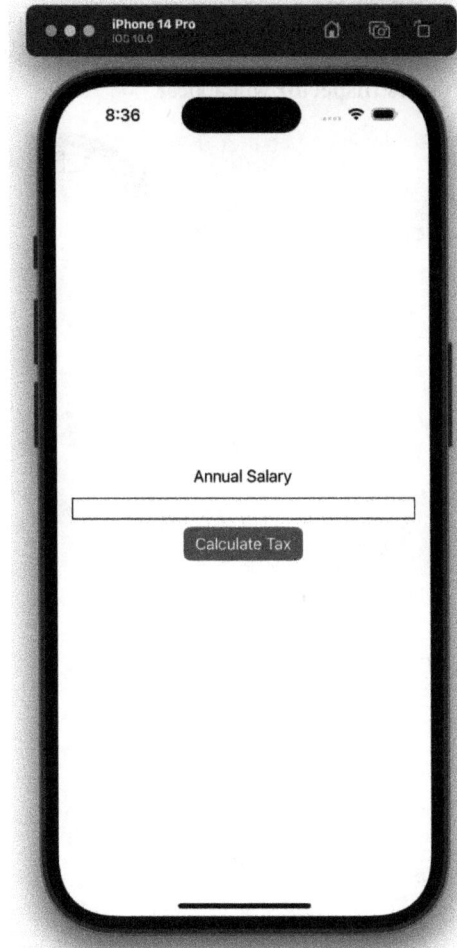

Figure 2.14 – Updated code preview

The preview shows we are very close. For the type of data we are inserting into TextField, it doesn't need to be this wide. Let's make it smaller. Modify TextField as follows:

```
TextField( "", text: $salary )
            .frame( width: 200.0 )
            .border( Color.black, width: 1 )
```

We have added a width of 200 to make the TextField look better suited for what we need. Thus far, we have changed the properties of our app's components programmatically. However, you can use the Xcode UI to tweak the properties as well. Doing this is simple: select a component in the code by hovering the mouse cursor over it and clicking the code as if you are going to edit it. Now, on the right, a pane including the **Attributes Inspector** will appear.

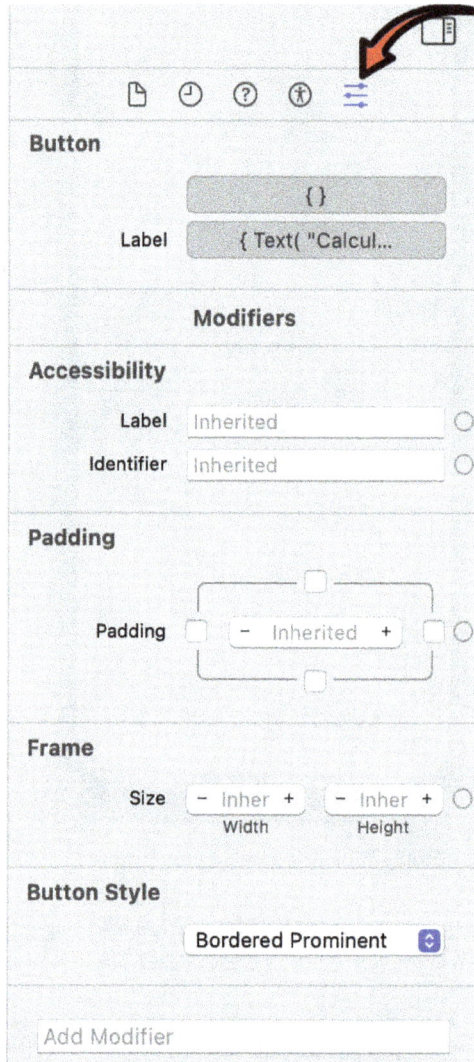

Figure 2.15 – Attributes Inspector

If the **Attributes Inspector** pane doesn't appear, go to **View | Inspectors | Attributes**.

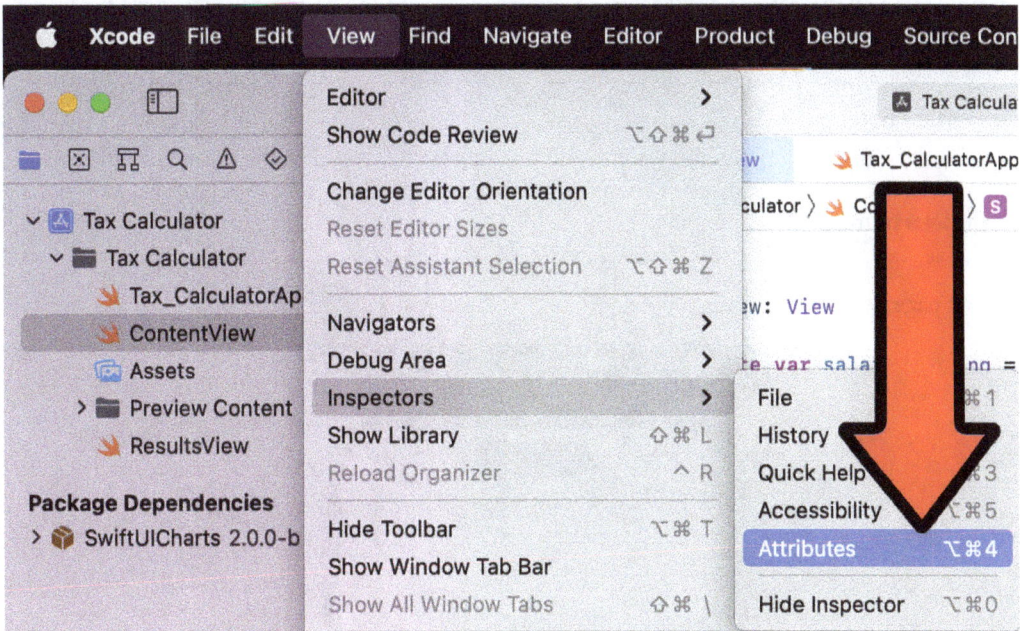

Figure 2.16 – Opening Attributes Inspector manually

We are almost done; we only have three UI components. The interface is currently quite compact. Let's spread out the components to make it look nicer. Add the following code to space out the components:

```
VStack
{
    Text( "Annual Salary" )
        .padding(.bottom, 75.0)

    TextField( "", text: $salary )
        .frame( width: 200.0 )
        .border( Color.black, width: 1 )
        .padding( .bottom, 75.0 )

    Button
    { }
    label:
    { Text( "Calculate Tax" ) }
        .buttonStyle( .borderedProminent )
}
.padding( )
```

In the preceding code, we added padding at the bottom of the `Text` and `TextField` components to evenly spread all three items out. Feel free to experiment with the padding values to get the UI to feel like what you are looking for.

> **Important note**
>
> If you have four components and want to add padding to the top three, the formula would be `n - 1` in terms of the number of components that need padding to be evenly spread out. `n` is the total number of components.

Our front page now looks like the following:

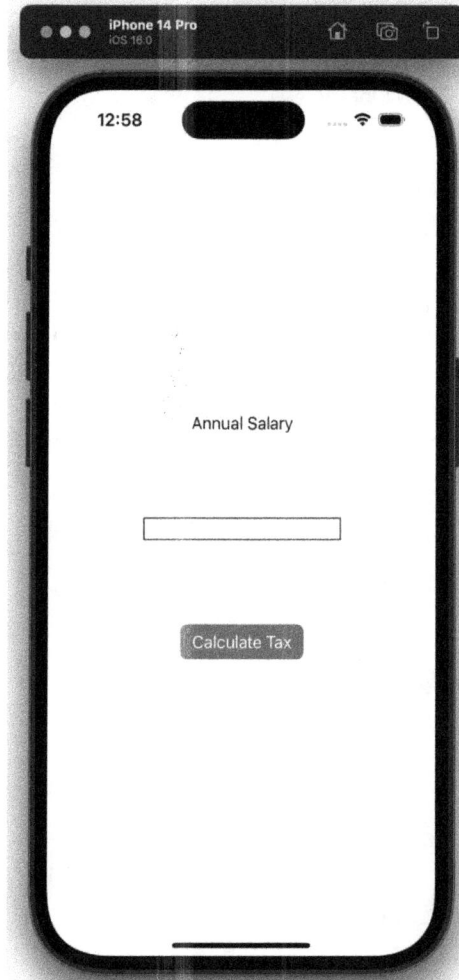

Figure 2.17 – Preview with padding

Right now, if we launch our app and click the `TextField`, a regular keyboard appears, as shown in the following screenshot:

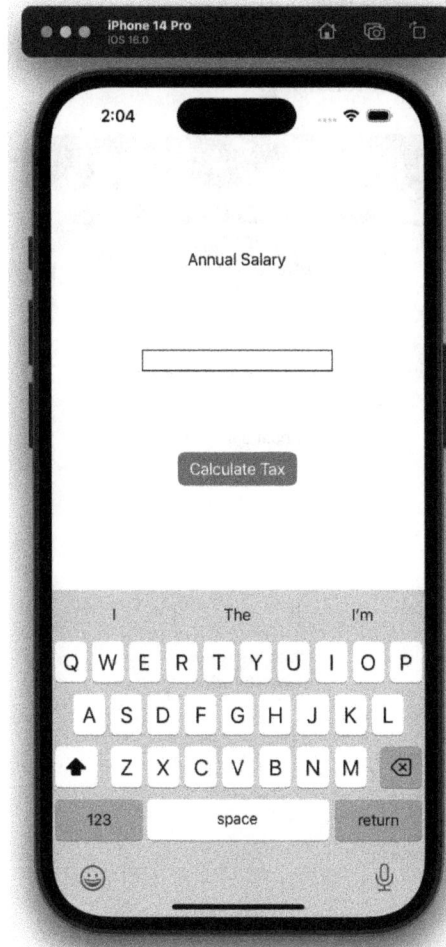

Figure 2.18 – Front page of the regular keyboard preview

This is fine for a field that requires text input for names or addresses, but this field only requires a salary and therefore only needs numerical input. Let's update our code to set the keyboard type to `decimalPad`:

```
TextField( "", text: $salary )
    .frame( width: 200.0 )
    .border( Color.black, width: 1 )
    .padding( .bottom, 75.0 )
    .keyboardType( .decimalPad )
```

> **Important note**
> There is a `numberPad` option but it doesn't allow the input of decimal numbers, so we will go ahead and use the `decimalPad` type.

If you run the app now, it will show the following:

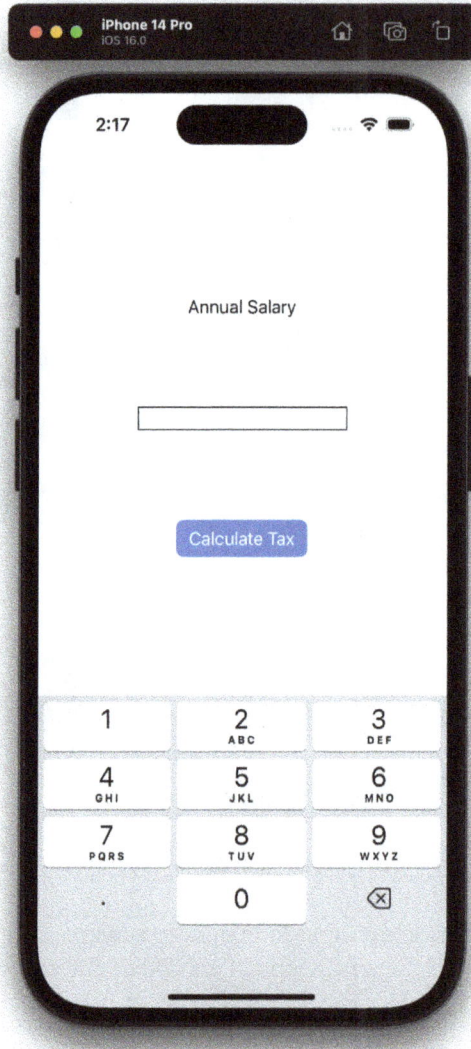

Figure 2.19 – Front page decimal pad preview

The keyboard type can also be changed in the **Attributes Inspector**. This is a great place to quickly see all the available keyboard types:

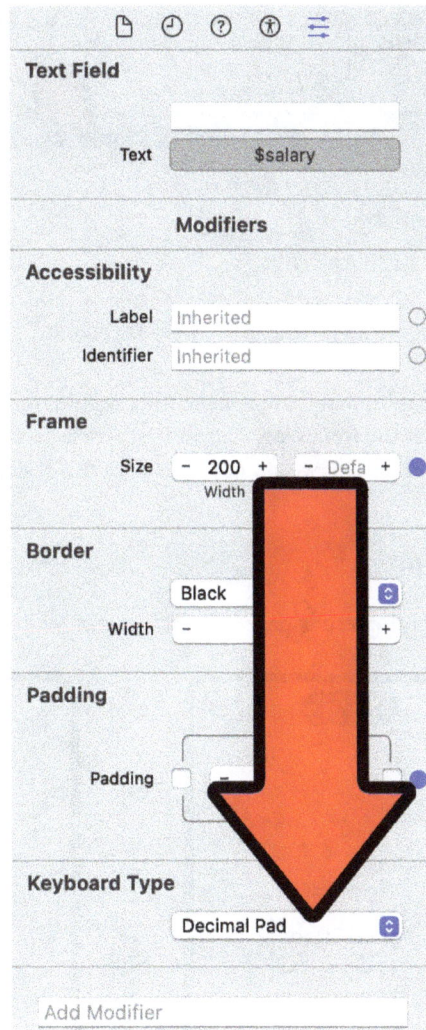

Figure 2.20 – Keyboard Type

> **Important note**
> For more information on keyboard types, check out https://developer.apple.com/
> documentation/swiftui/view/keyboardtype(_:).

If the keyboard doesn't show in the simulator, this is due to the fact your Mac already has a keyboard and the simulator decides you don't need it displayed. But this can be overridden. Either use the ⌘ + K keyboard shortcut to open and ⌘ + ⇧ + K to close it, or go to **I/O | Keyboard | Toggle Software Keyboard**:

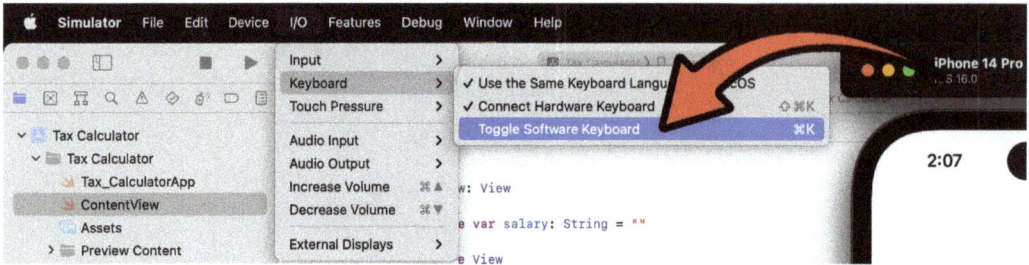

Figure 2.21 – Toggle Software Keyboard

Now the software keyboard in the simulator will appear. This should only need to be done once. We have now completed the design for the front page. Currently, there is no functionality, but this will be implemented in the following chapter. But we are not done with the design. We will now implement the design for the results page.

Implementing the results page

In this section, we will implement the results page's UI. As a reminder, here is what it will look like:

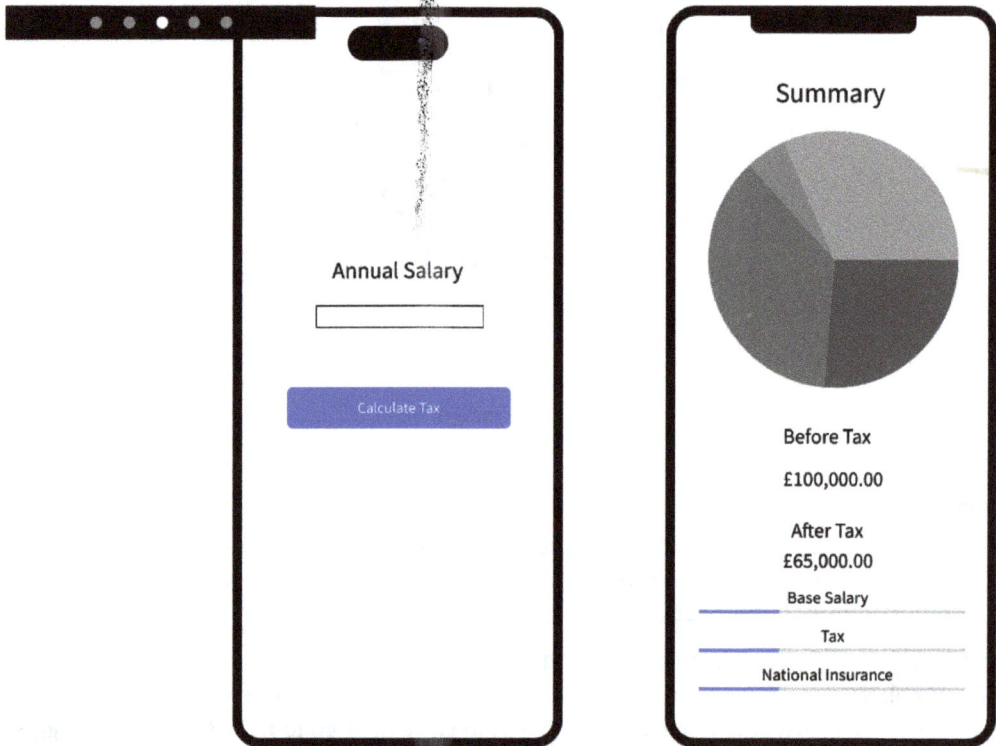

Figure 2.22 – Results page wireframe preview

There are three main sections on the results page. Each section is composed of two or more components. As a little task, see whether you can figure out what they are. Don't worry if you don't know the exact UI component names as we will take a look at them in the following sections.

Graph Summary Section

The graph summary section comprises two main components, a Text component and a pie chart. SwiftUI doesn't provide a pie chart, so we will use an external library. We will use the `ChartView` library created by *Andras Samu*, which can be found here: `https://github.com/AppPear/ChartView`.

This section will visually showcase a simple breakdown of the tax calculation.

Summary

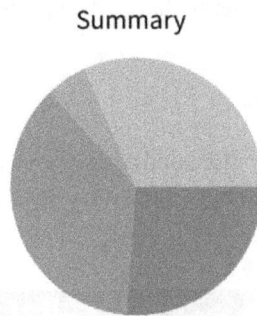

Figure 2.23 – Graph summary wireframe

Text Summary Section

In the text summary section, there are four text components. The first component informs the user that the following `Text` component is used to display the **Before Tax** salary title. The second component tells the user that the following text component is used to display the **After Tax** salary title. This does not include a breakdown of how the tax is split. This will come in the next section:

Before Tax

£100,000.00

After Tax
£65,000.00

Figure 2.24 – Text summary wireframe

Individual breakdown section

The individual breakdown section displays how the tax and salary are broken down. There are six components, three `Text` components and three `ProgressView` components. Each is paired together to make three subsections, **Base Salary**, **Tax**, and **National Insurance**. This design is simple but extendible. Once it is created, I give you the task of adding further breakdowns of the tax, such as student loans and pension:

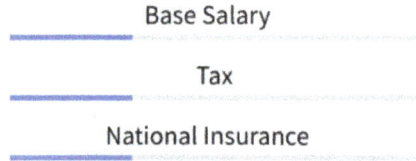

Figure 2.25 – Individual breakdown wireframe

In the next section, we will add the elements that make up the results page before wrapping up this chapter.

Adding results page components

In this section, we will add the previously discussed components to our results page. However, firstly we must integrate the `ChartView` framework by *Andras Samu*. Follow these steps:

1.   Go to **File | Add Packages…**:

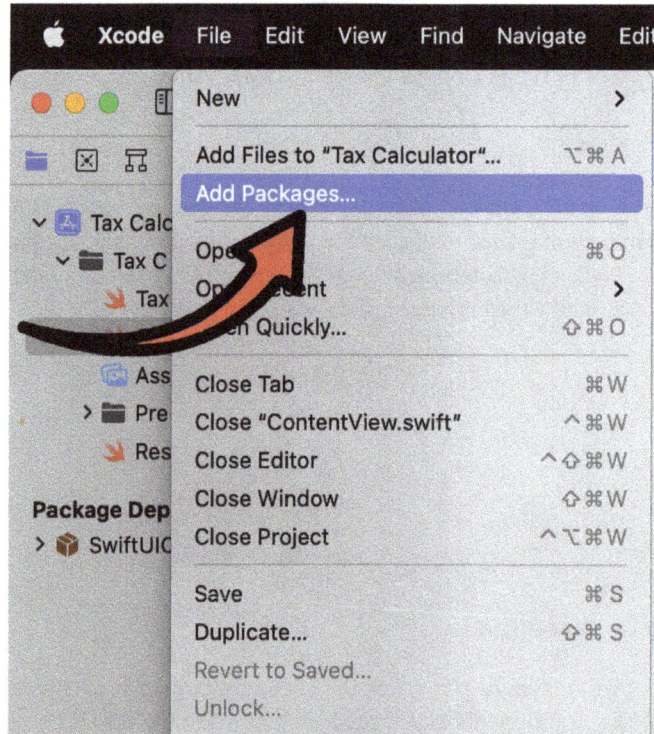

Figure 2.26 – Xcode Add Packages… option

2.   Search for the `ChartView` framework using the following URL: `https://github.com/AppPear/ChartView`.

3.   Select **Exact Version** from the **Dependency Rule** dropdown and set the text input to `2.0.0-beta.2`, or whatever the latest version is for you. Then, click **Add Package** at the bottom right. It is grayed out in mine as I have already added it:

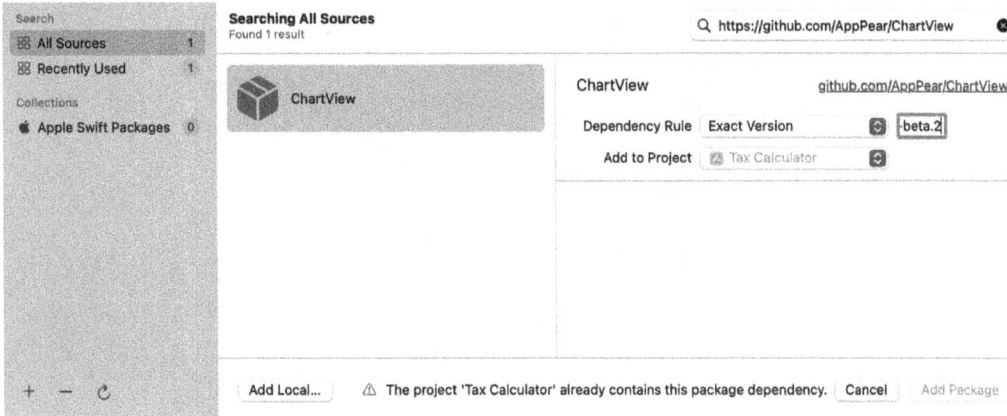

Figure 2.27 – Search for the ChartView package

4.   Clicking on **Add Package** will add the package to the project successfully.

5.   We will now create a new SwiftUI View for the results page. Right-click the calculator folder inside of your **Project Navigator** pane and select **New File…**:

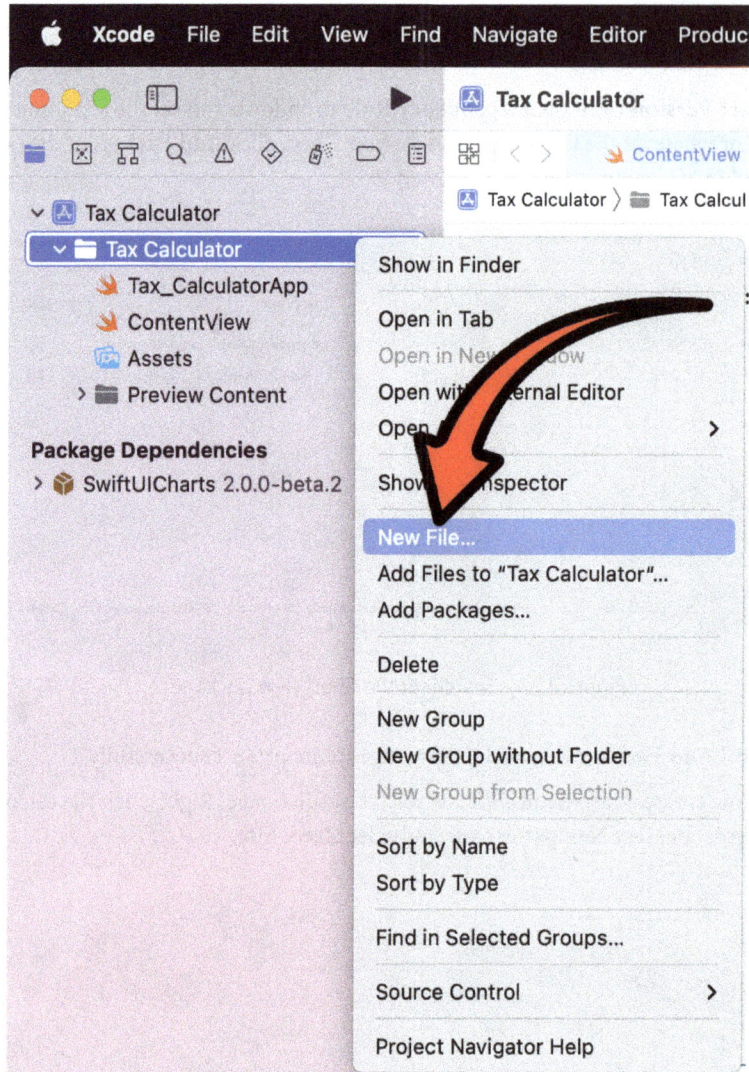

Figure 2.28 – New File…

6. Next, we will select the type of file we want to add, which for us is a SwiftUI View (selecting this provides a SwiftUI template, which saves us time and effort retyping the SwiftUI file structure every time), under the **User Interface** section:

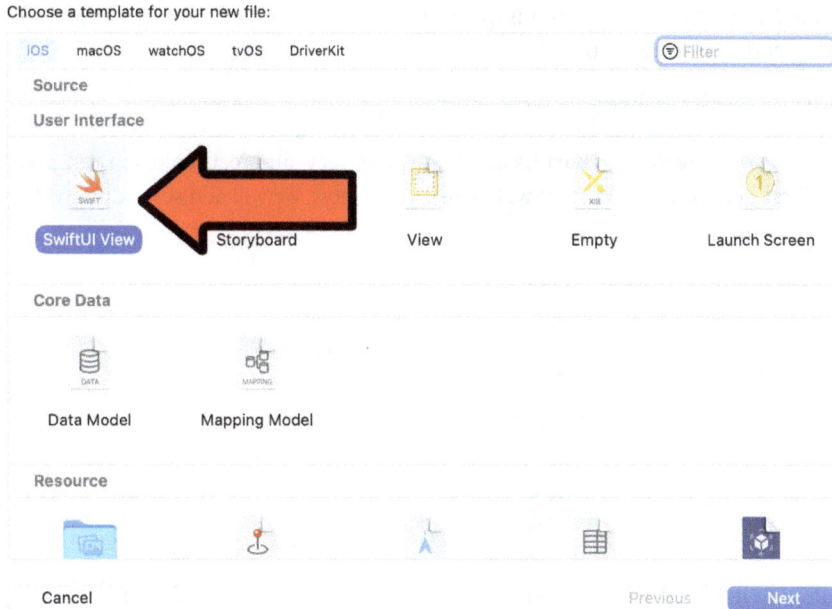

Figure 2.29 – SwiftUI View selection

7.   Finally, we must rename our **SwiftUI View**. Let's name it `ResultsView` and press **Create**:

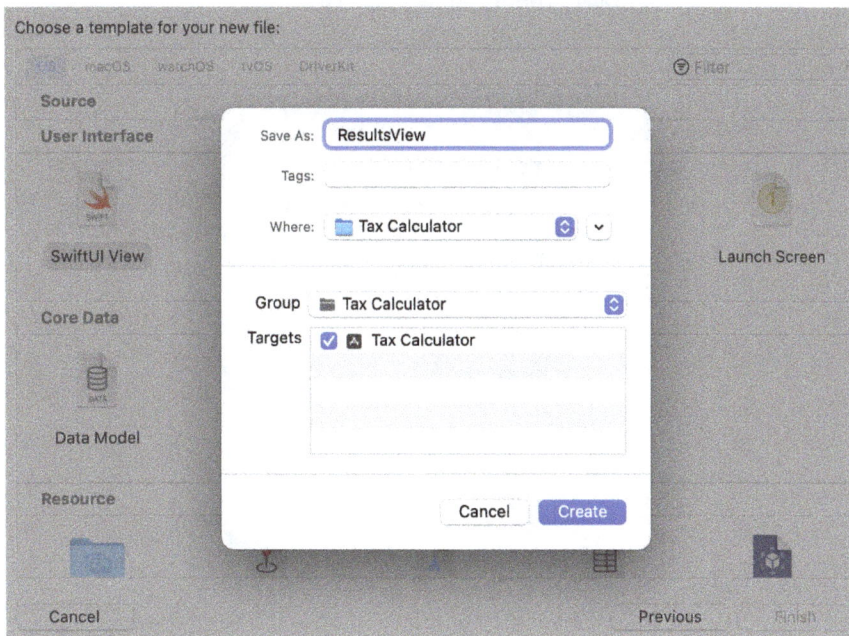

Figure 2.30 – View naming

8. Open the `ResultsView` file and import the `SwiftUICharts` framework by adding the following code to the top of the file:

```
import SwiftUICharts
```

9. Next, we need to create the chart itself. Doing so is very simple, thanks to the `ChartView` library. First, add the data the chart will be using. For now, we will add some dummy hardcoded data for testing:

```
struct ResultsView: View
{
    var taxBreakdown: [Double] = [5, 10, 15]


    var body: some View
    {

    }
}
```

The values in the array will represent the base salary, tax, and national insurance.

10. Next, we will implement our pie chart using `ChartView`:

```
struct ResultsView: View
{
    var taxBreakdown: [Double] = [5, 10, 15]

    var body: some View
    {
        PieChart( )
            .data( taxBreakdown )
            .chartStyle( ChartStyle( backgroundColor: .white,
                                     foregroundColor:
ColorGradient( .blue, .purple) ) )
    }
}
```

The preceding code will result in the following:

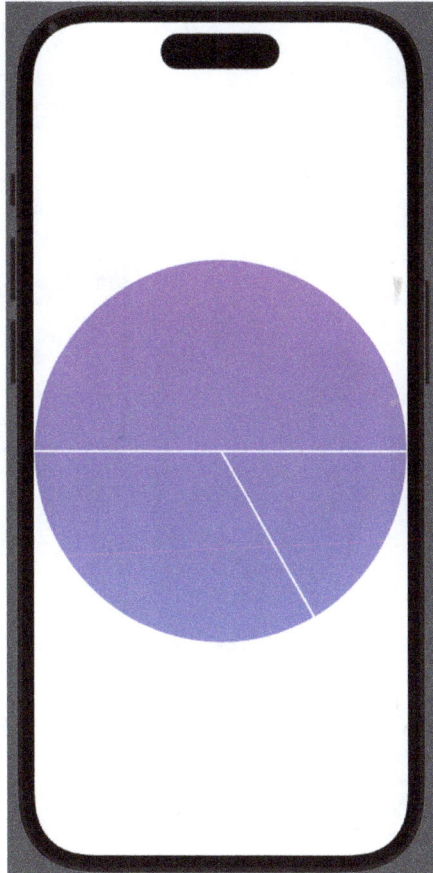

Figure 2.31 – Pie chart added

11. To view the `ResultsView`, you will need to use the **Live Preview Window**. By default, it should appear. If it doesn't, use the following keyboard shortcut: ⌥ + ⌘ + *Return*. Now, Xcode

will look like the following screenshot:

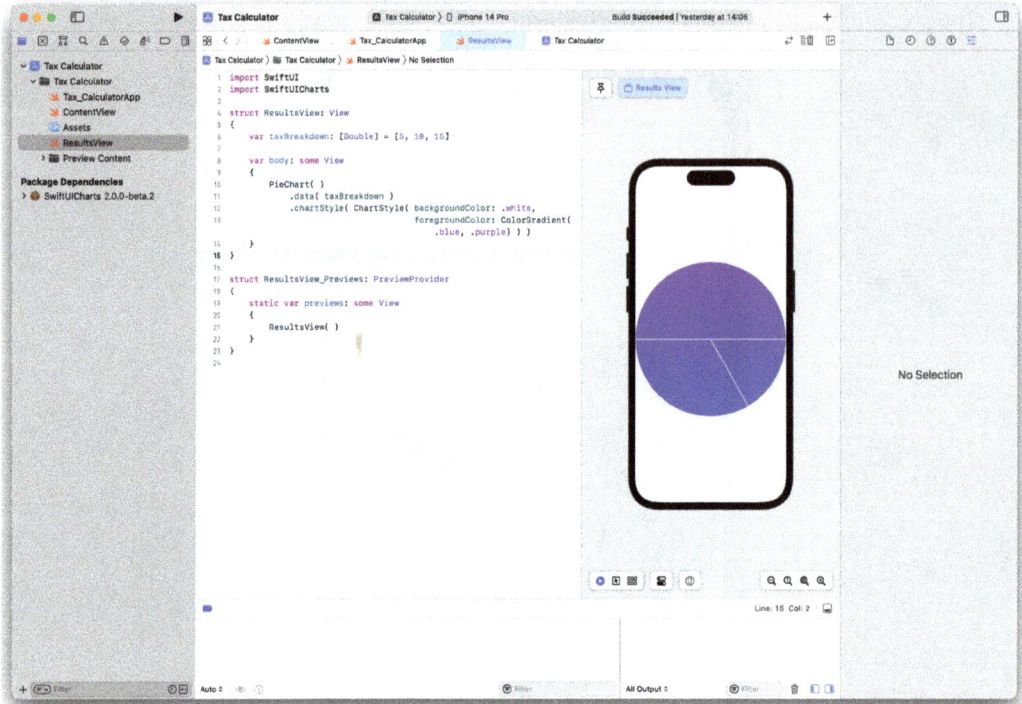

Figure 2.32 – Live preview window location

12. Right now, the pie chart goes up to the edges. Let's put the pie chart inside a VStack with padding. Edit the code as follows:

```
struct ResultsView: View
{
    var taxBreakdown: [Double] = [5, 10, 15]

    var body: some View
    {
        VStack
        {
            PieChart( )
                .data( taxBreakdown )
                .chartStyle( ChartStyle( backgroundColor:
.white,
                                         foregroundColor:
ColorGradient( .blue, .purple ) ) )
        }.padding( )
    }
}
```

The changes in the preceding code will now make the chart look like the following:

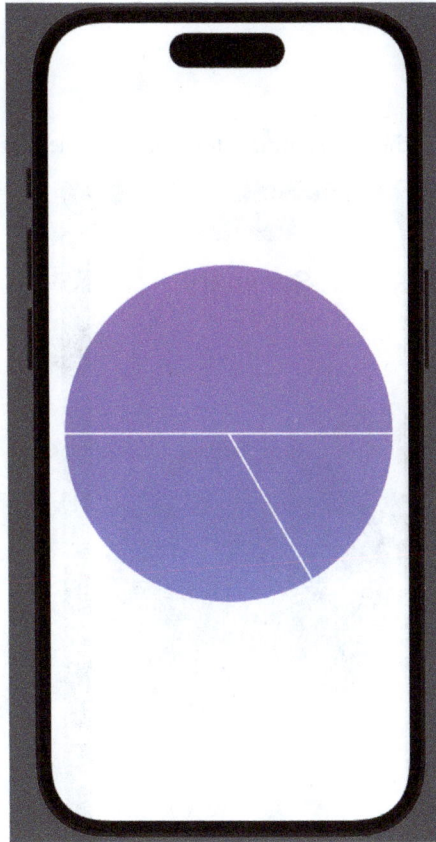

Figure 2.33 – Pie chart with padding

13. Let's add a `Text` component that says **Summary** above the pie chart. But as it will be a header, we will make it bold and set the font size to `36`. Add the following code to the body:

```
var body: some View
{
    VStack
    {
        Text( "Summary" )
            .font( .system( size: 36 ) )
            .fontWeight( .bold )

        PieChart( )
            .data( taxBreakdown )
            .chartStyle( ChartStyle( backgroundColor: .white,
```

```
                                              foregroundColor:
      ColorGradient(
      .blue,  .purple ) ) )
          }.padding( )
      }
```

The following output shows the new summary text above the pie chart that we added previously.

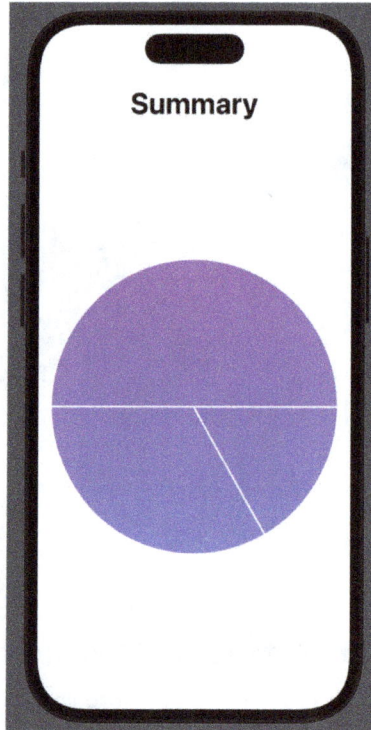

Figure 2.34 – Summary header

14. Below the pie chart, we will add four more text components, for **Before Tax** and **After Tax**: one for the heading of each subsection and one for the actual figure. For now, we will hardcode the values. Update the code as follows:

```
var body: some View
{
    VStack
    {
        Text( "Summary" )
            .font( .system( size: 36 ) )
            .fontWeight( .bold )
```

```
        PieChart ( )
            .data ( taxBreakdown )
            .chartStyle ( ChartStyle ( backgroundColor: .white,
                                       foregroundColor:
ColorGradient (
.blue, .purple ) ) )

        Text ( "Before Tax" )
            .font ( .system ( size: 32 ) )

        Text ( "£100,000.00" )
            .font ( .system ( size: 32 ) )

        Text ( "After Tax" )
            .font ( .system ( size: 32 ) )

        Text ( "£65,000.00" )
            .font ( .system ( size: 32 ) )
    }.padding ( )
}
```

The preceding code will display the following:

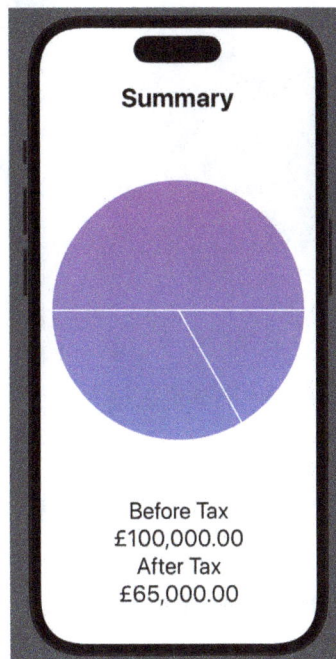

Figure 2.35 – Before Tax and After Tax text

15. Right now, the text components on the bottom of the page are crammed together. Let's add padding to the top and bottom of each text component to spread them out. You can obviously use the **Attributes Inspector** to do this, but we will do it programmatically:

```
Text( "Before Tax" )
    .font( .system( size: 32 ) )
    .padding(.vertical)
Text( "£100,000.00" )
    .font( .system( size: 32 ) )
    .padding(.vertical)
Text( "After Tax" )
    .font( .system( size: 32 ) )
    .padding(.vertical)
Text( "£65,000.00" )
    .font( .system( size: 32 ) )
    .padding(.vertical)
```

After adding the padding in the preceding code, we will have a results page that looks like this:

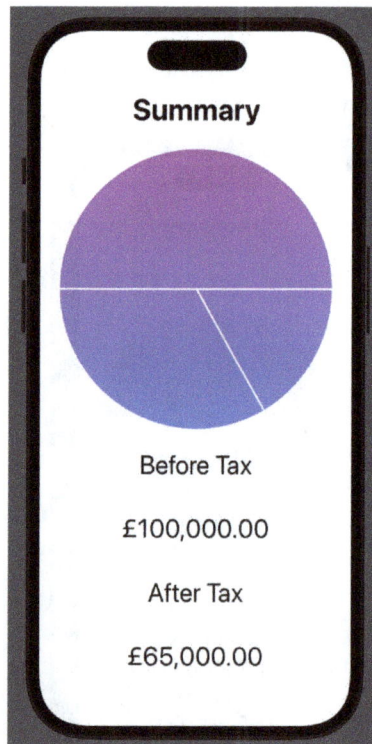

Figure 2.36 – Results after padding

16. Next, we need to add the progress bars, which will represent the salary, tax, and national insurance. We will use the `ProgressView` component combined with a `Text` component to display the tax breakdown. After the previously added `Text` components, add the following code:

```
Text( "Post Tax Salary" )

ProgressView( "", value: 20, total: 100 )

Text( "Tax" )

ProgressView( "", value: 20, total: 100 )
```

The preceding code adds the two `ProgressView` and `Text` component pairs, which shows the post-tax salary and tax. This will result in the following:

Figure 2.37 – Breakdown components added

17. You have probably noticed that we only added two of the three `ProgressView` components. The reason for this is to showcase an error that occurs. So, now add the following code after the previously added code:

```
Text ( "National Insurance" )

ProgressView ( "", value: 20, total: 100 )
```

This will result in the following error: **Extra arguments at positions #11, #12 in call**. This is because you cannot have more than 10 components displayed. But there is space, so how do we get around this? The solution is simple. The limit of 10 is for components, not components within components. Put basically, we can group multiple components using the Group component, which will make the view detect it as one.

18. We will group the three ProgressView and Text components as follows:

```
Group
{
    Text ( "Post Tax Salary" )

    ProgressView ( "", value: 20, total: 100 )

    Text ( "Tax" )

    ProgressView ( "", value: 20, total: 100 )

    Text ( "National Insurance" )

    ProgressView ( "", value: 20, total: 100 )
}
```

This will solve the annoying 10-limit error and result in the following:

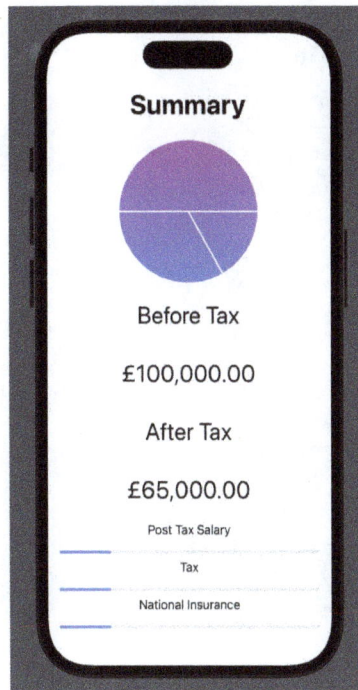

Figure 2.38 – Components grouped

Here is a look at the whole code now that we are finished with this section:

```swift
import SwiftUI
import SwiftUICharts

struct ResultsView: View
{
    var taxBreakdown: [Double] = [5, 10, 15]

    var body: some View
    {
        VStack
        {
            Text( "Summary" )
                .font( .system( size: 36 ) )
                .fontWeight( .bold )

            PieChart( )
                .data( taxBreakdown )
                .chartStyle( ChartStyle( backgroundColor: .white,
```

```
                                         foregroundColor:
ColorGradient( .blue, .purple ) ) )

        Text( "Before Tax" )
            .font( .system( size: 32 ) )
            .padding(.vertical)

        Text( "£100,000.00" )
            .font( .system( size: 32 ) )
            .padding(.vertical)

        Text( "After Tax" )
            .font( .system( size: 32 ) )
            .padding(.vertical)

        Text( "£65,000.00" )
            .font( .system( size: 32 ) )
            .padding(.vertical)

        Group
        {
            Text( "Post Tax Salary" )

            ProgressView( "", value: 20, total: 100 )

            Text( "Tax" )

            ProgressView( "", value: 20, total: 100 )

            Text( "National Insurance" )

            ProgressView( "", value: 20, total: 100 )
        }
    }.padding( )
    }
}

struct ResultsView_Previews: PreviewProvider
{
    static var previews: some View
    {
        ResultsView( )
    }
}
```

We have covered a lot in this vast section. We started off by adding an external framework, which we saw is very easy to integrate and extremely powerful. The framework allowed us to easily implement a pie chart. This is very useful as not all basic features are provided by Apple in Swift and SwiftUI, so being able to add external code bases makes the development process less painful. After that, we implemented the pie chart, text summaries, and progress views to further illustrate the tax breakdown.

Summary

In this chapter, we covered the design of our tax calculator application. We looked at wireframes and broke down each element into SwiftUI components. We then implemented the SwiftUI components to match the design from the wireframes. We also took a look at the requirements for building this application, and the design specifications, which looked at the features a tax calculator app can have. Then, we simplified it to the core features our app will provide. We further advanced in the design specifications with acceptance criteria for what we would like our app to do. We also looked at how external libraries can be integrated to provide additional functionality.

In the next chapter, we'll take a look at implementing the tax calculation backend functionality and tying the two views together.

3
iPhone Project – Tax Calculator Functionality

In this chapter, we will work on implementing the tax calculation and page navigation functionality in our first project, the tax calculator. In the previous chapter, we looked at the design of the calculator and broke it down into two views and all the components required. We then implemented all the components using SwiftUI. At the end of the previous chapter, we effectively only had a fancy wireframe. Now, we will implement all the functionality to provide navigation between the two views, calculating the tax breakdown and displaying the calculation.

This chapter will be split into the following sections:

- Navigating from `ContentView` to `ResultsView`
- Input validation
- Calculating tax breakdown
- Extra tasks

By the end of this chapter, you will have created a fully functional tax calculator that can be used as a foundation. I'll provide exercises as we reach the end of the chapter to implement more advanced functionality within the tax calculator. The code is yours to use, modify, and distribute as you see fit. This will transition nicely into our next project, the iPad gallery app.

Technical Requirements

This chapter requires you to download Xcode version 14 or above from Apple's App Store.

To install Xcode, just search for `Xcode` in the App Store, and select and download the latest version. Open Xcode and follow any additional installation instructions. Once Xcode has opened and launched, you're ready to go.

Version 14 of Xcode has the following features/requirements:

- Includes SDKs for iOS 16, iPadOS 16, macOS 12.3, tvOS 16, and watchOS 9

- Supports on-device debugging in iOS 11 or later, tvOS 11 or later, and watchOS 4 or later

- Requires a Mac running macOS Monterey 12.5 or later

- For further information regarding the technical details, please refer to *Chapter 1*.

The code files for this chapter and the previous chapter to use as a base can be found here: `https://github.com/PacktPublishing/Elevate-SwiftUI-Skills-by-Building-Projects`

Navigating from ContentView to ResultsView

In this section, we will finally implement the navigational system for moving from `ContentView` to `ResultsView` and back again.

If you recall back to the previous chapter, when viewing the UI for the `ResultsView`, we were forced to use the Live Preview Window instead of running the application. The reason was that there was no functionality for navigating from `ContentView` to `ResultsView`. We have already added the button for calculating the tax, but we need to implement the code for the button triggering the navigation.

First, we need to wrap our `VStack` in `ContentView` in a `NavigationView`. The `NavigationView` allows us to present a stack of views which is very useful for navigation, as it effectively has a history of all previous views, allowing an easy and extendable navigation system:

```swift
import SwiftUI
struct ContentView: View
{
    @State private var salary: String = ""

    var body: some View
    {
        NavigationView
        {
            VStack
            {
                Text( "Annual Salary" )
                    .padding(.bottom, 75.0)

                TextField( "", text: $salary )
                    .frame( width: 200.0 )
                    .border( Color.black, width: 1 )
                    .padding( .bottom, 75.0 )
                    .keyboardType( .decimalPad )
```

```
                Button
                { }
                label:
                { Text( "Calculate Tax" ) }
                    .buttonStyle( .borderedProminent )
            }
          .padding( )
        }
    }

    func GoToResultsView( )
    {
        ResultsView( )

    }
}

struct ContentView_Previews: PreviewProvider
{
    static var previews: some View
    {
        ContentView( )
    }
}
```

We have now implemented `NavigationView`, which allows us to use our existing code and design for navigation. There is one extra piece of code we require – a `GoToResultsView` function. An empty version has been added to the preceding code. We will use it later in this chapter.

As of now, this won't have any effect on our application, as we need to modify the button to become a `NavigationLink`, which is basically a fancy button allowing us to navigate between views. Fortunately for us, it can be styled in any way we desire, but first let's implement a basic `NavigationLink`. To do this, replace the button code in `ContentView` with the following code:

```
NavigationLink( destination: ResultsView( ), label:
{
    Text( "Calculate Tax" )
} )
```

Let's break down the code we just added. The first parameter it takes is the destination for the application's view system, so the new view that will be pushed onto the stack. We have specified `ResultsView` but you could easily specify a simple component, which would create its own view. The next parameter is called `label`; this is similar to the label specified in the button itself. Currently, we have a `Text` component, which looks like this:

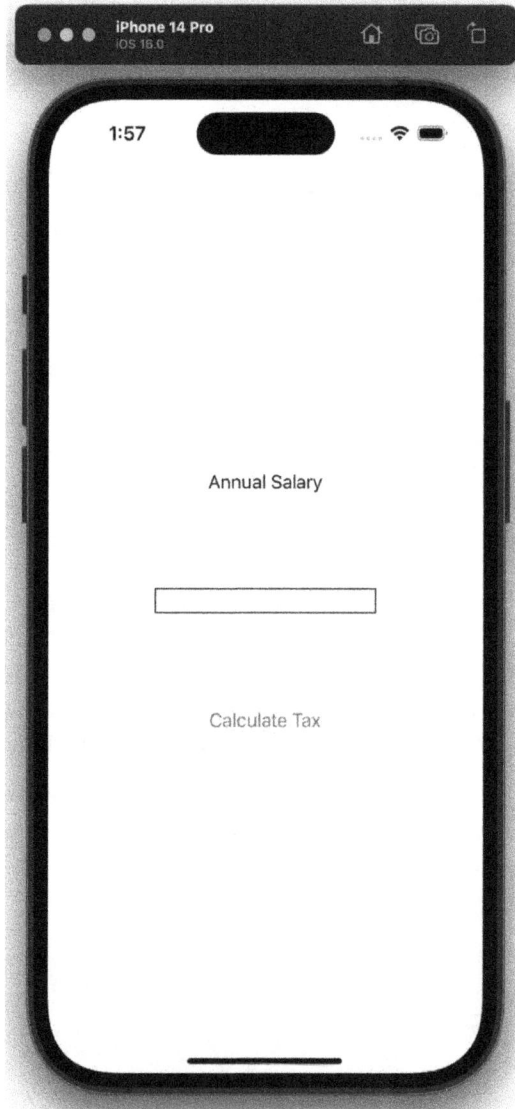

Figure 3.1 – Basic NavigationLink button

Let's style the `NavigationLink` button and modify the code to look as follows:

```
NavigationLink( destination: ResultsView( ), label:
{
    Text( "Calculate Tax" )
        .bold( )
        .frame( width: 200, height: 50 )
        .background( Color.blue )
        .foregroundColor( Color.white )
        .cornerRadius( 10 )
} )
```

We have added five main style aspects to the button. Let's take a look at each one:

- `.bold( )`: Makes the text bold
- `.frame( width: 200, height: 50 )`: Sets the size of the button
- `.background( Color.blue )`: Sets the background color to blue
- `.foregroundColor( Color.white )`: Sets the text color to white
- `.cornerRadius( 10 )`: Makes the button's corners rounded, giving a more natural, iOS-like feel

After these styles have been applied to the application and, more specifically, the button, it will look like the following figure:

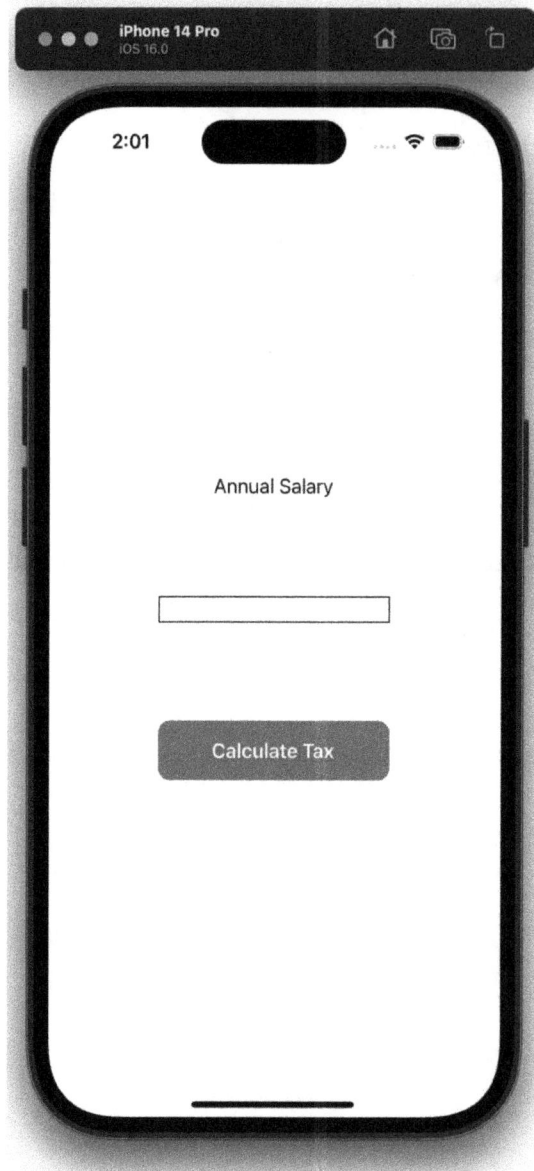

Figure 3.2 – NavigationLink button styled

We will now add a navigation title. This provides a nice consistent method for adding a title/header to the view. This is simple – add the following code to the bottom of the VStack, just after the padding:

```
.padding( )
.navigationTitle( "Main Page" )
```

Running the application will show the following:

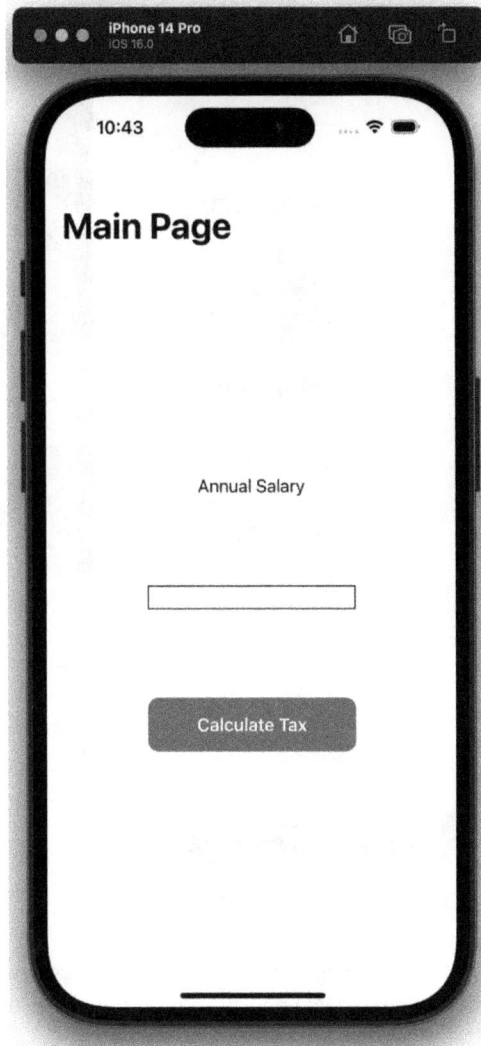

Figure 3.3 – Navigation title

Navigating to ResultsView by clicking the **Calculate Tax** button will show that the back button's text in the top left is now the same as the navigation title that was added. This can be seen in the following screenshot:

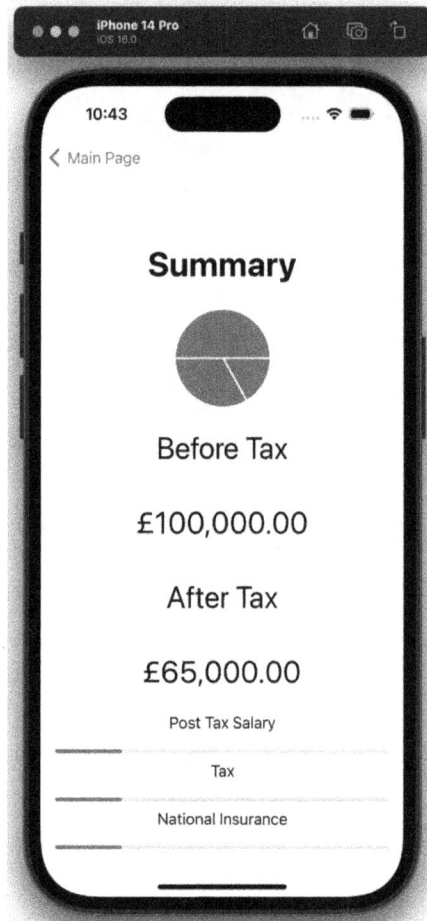

Figure 3.4 – Updated back button text

Let's now update ResultsView to use a navigation title instead of a Text component for the heading of the page. First, we need to remove the Text component with the following code from ResultsView:

```
Text( "Summary" )
    .font( .system( size: 36 ) )
    .fontWeight( .bold )
```

Now that ResultsView has no title, let's add the navigation link. Like before, add the following code after the VStack, right after the padding:

```
.padding( )
.navigationBarTitle( "Summary" )
```

This little addition will change `ResultsView` as follows:

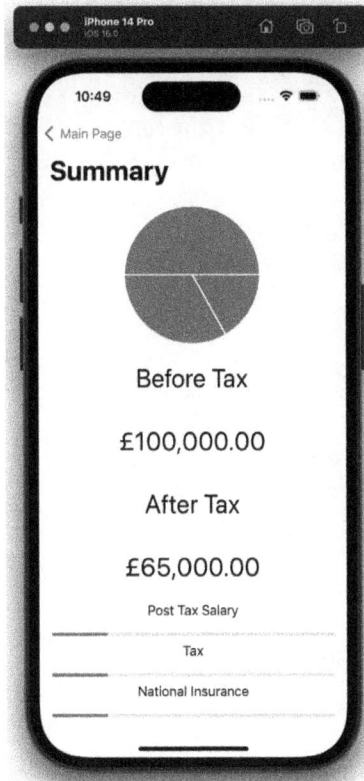

Figure 3.5 – ResultsView navigation title

So, we have added a fair bit of code. This is what the `ContentView` and `ResultsView` code should look like before we move on:

ContentView

```
import SwiftUI
struct ContentView: View
{
    @State private var salary: String = ""
    var body: some View
    {
        NavigationView
        {
            VStack
```

```
            {
                Text( "Annual Salary" )
                    .padding(.bottom, 75.0)

                TextField( "", text: $salary )
                    .frame( width: 200.0 )
                    .border( Color.black, width: 1 )
                    .padding( .bottom, 75.0 )
                    .keyboardType( .decimalPad )

                NavigationLink( destination: ResultsView( ), label:
                {
                    Text( "Calculate Tax" )
                        .bold( )
                        .frame( width: 200, height: 50 )
                        .background( Color.blue )
                        .foregroundColor( Color.white )
                        .cornerRadius( 10 )
                } )
            }
            .padding( )
            .navigationTitle( "Main Page" )
        }
    }
}
struct ContentView_Previews: PreviewProvider
{
    static var previews: some View
    {
        ContentView( )
    }
}
```

ResultsView

```
import SwiftUI
import SwiftUICharts
struct ResultsView: View
{
    var taxBreakdown: [Double] = [5, 10, 15]
    var body: some View
    {
        VStack
```

```
        {
            PieChart( )
                .data( taxBreakdown )
                .chartStyle( ChartStyle( backgroundColor: .white,
                                        foregroundColor:
ColorGradient( .blue, .purple ) ) )

            Text( "Before Tax" )
                .font( .system( size: 32 ) )
                .padding(.vertical)

            Text( "£100,000.00" )
                .font( .system( size: 32 ) )
                .padding(.vertical)

            Text( "After Tax" )
                .font( .system( size: 32 ) )
                .padding(.vertical)

            Text( "£65,000.00" )
                .font( .system( size: 32 ) )
                .padding(.vertical)

            Group
            {
                Text( "Post Tax Salary" )

                ProgressView( "", value: 20, total: 100 )

                Text( "Tax" )

                ProgressView( "", value: 20, total: 100 )

                Text( "National Insurance" )

                ProgressView( "", value: 20, total: 100 )
            }
        }
        .padding( )
        .navigationBarTitle( "Summary" )
    }
}
struct ResultsView_Previews: PreviewProvider
{
```

```
    static var previews: some View
    {
        ResultsView( )
    }
}
```

Wow, we have done a lot. Take the time to pat yourself on the back. All this allowed us to implement a seamless yet familiar navigation system. We learned how to implement a navigation system to navigate to `ResultsView` and back to `ContentView` easily.

In the next section, we will validate the salary to make sure we do not pass through any data that is incorrect.

Validating salary input

As of now, if you press the **Calculate Tax** button, it takes the user to the results page from the front page. However, it does this regardless of input, so it will go to the next page even if there is no salary. The following validation checks must be done for it to be an acceptable value:

- Does it contain a value that is a number?

- Is the value above 0 (this rules out negatives)?

You might be wondering why we can't just check whether the salary is above 0 as we have chosen a decimal keypad. There are two main reasons for this:

- The user can insert decimal points in a way that makes the input **Not a Number** (**NaN**) – for example, `4.5.6...`

- Even though the user cannot directly type text due to the keyboard being a decimal pad, they can still copy it from another application and paste it into our calculator, thus breaking the number-only `TextField`. You may think it's worth just disabling pasting but it's important to retain this functionality, as the user may legitimately want to paste a number instead of typing it out, especially if it isn't a simple small number.

If not correctly handled, the application will crash upon the user pressing the **Calculate Tax** button. The way we will achieve these conditions is explained in the following sections.

Using a variable to track if the salary is valid

We will add a Boolean variable after the salary string to track whether the salary is valid; if it is, then `ResultsView` will be displayed. The salary string is only checked when the **Calculate Tax** button is pressed. Add the following code before the body:

```
@State private var isSalaryValid: Bool = false
```

isActive NavigationLink

Now we will link the isSalaryValid variable created previously with NavigationLink using the isActive parameter. Update the NavigationLink code to look like the following:

```
NavigationLink( destination: ResultsView( ), isActive: $isSalaryValid,
label:
```

isActive is a simple concept – if true, it immediately takes you to the destination view, and if it is false, then nothing happens.

In the following section, we will validate the salary.

Validating the salary

Firstly, we will override the tap gesture and call a custom function called GoToResultsView. Add the following code to the end of all text properties:

```
.onTapGesture
{
    GoToResultsView( )
}
```

Finally, we must check whether the string is valid, then navigate to the results view. Add the following function to handle this:

```
func GoToResultsView( )
{
    if ( Float( salary ) != nil )
    {
        if ( Float( salary )! > 0 )
        { isSalaryValid = true }
    }
}
```

Let's unpack this function to see what it does:

- if (Float(salary) != nil): Checks whether the salary is a number. This is accomplished by casting the string to a Float. If unsuccessful, it is nil – this allows us to check for this. This also includes validating an empty string.

- if (Float(salary)! > 0): Checks whether the salary is greater than zero. The exclamation mark says that the variable salary is definitely a Float to avoid any problems, as it's already been checked we can do this.

- `isSalaryValid` = `true`: Sets the `isSalaryValid` check variable to `true`, which triggers `ResultsView` and loads it up.

All of these additions should result in the following code in the following `ContentView` file:

```
import SwiftUI
struct ContentView: View
{
    @State private var salary: String = ""
    @State private var isSalaryValid: Bool = false
    var body: some View
    {
        NavigationView
        {
            VStack
            {
                Text( "Annual Salary" )
                    .padding(.bottom, 75.0)

                TextField( "", text: $salary )
                    .frame( width: 200.0 )
                    .border( Color.black, width: 1 )
                    .padding( .bottom, 75.0 )
                    .keyboardType( .decimalPad )

                NavigationLink( destination: ResultsView( ), isActive:
$isSalaryValid, label:
                    {
                    Text( "Calculate Tax" )
                        .bold( )
                        .frame( width: 200, height: 50 )
                        .background( Color.blue )
                        .foregroundColor( Color.white )
                        .cornerRadius( 10 )
                        .onTapGesture
                        {
                            GoToResultsView( )
                        }
                    } )
            }
            .padding( )
            .navigationTitle( "Main Page" )
        }
    }
}
```

```
    func GoToResultsView( )
    {
        if ( Float( salary ) != nil )
        {
            if ( Float( salary )! > 0 )
            { isSalaryValid = true }
        }
    }
}
struct ContentView_Previews: PreviewProvider
{
    static var previews: some View
    {
        ContentView( )
    }
}
```

With the preceding code, we validated the salary, and a variable was used to track the status of the validation, which allowed us to navigate to ResultsView when it was validated.

In the next section, we will pass the salary between ContentView and ResultsView.

Passing the salary through to ResultsView

As of now, when we click **Calculate Tax**, ResultsView is still using dummy data. Let's change that by passing in the salary we validated in the previous section. Doing this is simple when using State and Bind. The salary variable in ContentView is already a state variable, simply meaning when it changes, the part of the view linked to it also changes and vice versa. When we change the text in the TextField salary, our application updates the state variable. We can use a binding variable in ResultsView, which allows us to pass in data between views.

First, in ResultsView, add the following line above the taxBreakdown array:

```
@Binding var salary: String
```

This simply declares a salary variable of type String, which is the same format as the salary variable in ContentView. @Binding just states that it is expecting this value to be passed in.

Now we must update NavigationLink to pass in the salary variable from ContentView to ResultsView like so:

```
NavigationLink( destination: ResultsView( salary: $salary ), isActive:
$isSalaryValid, label:
```

Though we have completed all the binding, if we try and run the application, the following error will be thrown:

```
static var previews: some View
{
    ResultsView( )        ● Missing argument for parameter 'salary' in call
}
```

Figure 3.6 – Preview error

This error relates to the preview view, which appears usually to the right of the code. Without this, the preview cannot run. To fix this error, we can give a default hardcoded value just for the preview. Update the code as follows:

```
struct ResultsView_Previews: PreviewProvider
{
    static var previews: some View
    {
        ResultsView( salary: .constant( "100" ) )
    }
}
```

Now once we run the application, we get no errors because we solved the missing argument error, and can move on to calculating the tax breakdown from the salary we passed through.

> **Note**
>
> If you would like to test to make sure the variable has been passed through, you can use a breakpoint or `print` statement. I will let you do that as an extra task.

Calculating tax breakdown

In the previous chapter, we passed through the `salary` variable, but we still need to calculate the tax. I will be doing it in line with the UK income tax rates of 2023/2024, but this can be adapted easily for any other tax system. In the following table, we have the tax rates for 2023/2024:

Income	Tax rate	
Up to £12,570	0%	Personal allowance
£12,571 to £37,700	20%	Basic rate
£37,701 to £150,000	40%	Higher rate
over £150,000	45%	Additional rate

Table 3.1 – Tax brackets

As there are brackets and not a single fixed tax, we will need to do a few calculations to figure out the exact tax to be deducted. On top of the income tax, there is also national insurance tax. National insurance tax isn't so simple as there are different categories, but let's keep it at a simple 13%, which it roughly is.

> **Note**
> There are definitely more aspects to calculating tax, such as pension contributions, student loans, and so on; however, we will leave it at this.

Tax calculation

Let's implement a formula in `ResultsView` to calculate the income tax. Add the following code at the start of the body:

```
var body: some View {
let salaryNum = Double( salary )!
var incomeTax: Double = 0

if ( salaryNum > 12570 )
{
    if ( salaryNum > 37700 )
    {
        if ( salaryNum > 150000 )
        {
            incomeTax += ( 37700 - 12571 ) * 0.2
            incomeTax += ( 150000 - 37701 ) * 0.4
            incomeTax += ( salaryNum - 150000 ) * 0.45
        }
        else
        {
            incomeTax += ( 37700 - 12571 ) * 0.2
            incomeTax += ( salaryNum - 37700 ) * 0.4
        }
    }
    else
    { incomeTax += ( salaryNum - 12570 ) * 0.2 }
}

return VStack {
.... .
}
}
```

We check each tax bracket and calculate the tax accordingly. Next, we will calculate the national insurance. First, add another double variable below `incomeTax`:

```
let salaryNum = Double( salary )!
var incomeTax: Double = 0
var nationalInsuranceTax: Double = 0
```

Now we can simply calculate it by multiplying the salary by `0.13` to get `13%`. Add the following code below the income tax calculation:

```
nationalInsuranceTax = salaryNum * 0.13
```

Now that we have all of the taxes calculated, we can move on to working out the post-tax salary. This is simple – we just subtract the income tax and national insurance tax from the salary. Add the following code beneath the previous code:

```
let postTaxSalary = salaryNum - incomeTax - nationalInsuranceTax
```

Let's format these values into strings, which will be used when displaying the tax figures. Add the following code below `post-tax salary` calculation:

```
let salaryString = String( format:"£%.2F", salaryNum )
let postTaxSalaryString = String( format: "£%.2F", postTaxSalary )
let incomeTaxString = String( format: "£%.2F", incomeTax )
let nationalInsuranceTaxSting = String( format: "£%.2F",
nationalInsuranceTax )
```

This creates a string from the respective numbers rounded to two decimal places. Next, move the `taxBreakdown` hardcoded pie chart array beneath the formatted strings, and update it as follows:

```
let taxBreakdown: [Double] = [postTaxSalary, incomeTax,
nationalInsuranceTax]
```

We are using let instead of var as it won't need to be changed.

Now we are finally ready to start updating each component in the UI. First, we will update the `Text` components, which display the `Before Tax` and `After Tax` figures like so:

```
Text( "Before Tax" )
    .font( .system( size: 32 ) )
    .padding(.vertical)

Text( salaryString )
    .font( .system( size: 32 ) )
    .padding(.vertical)

Text( "After Tax" )
    .font( .system( size: 32 ) )
    .padding(.vertical)

Text( postTaxSalaryString )
    .font( .system( size: 32 ) )
    .padding(.vertical)
```

Finally, update `ProgressView` to display the correct percentage and value for the tax and salary:

```
Text( "Post Tax Salary" )
ProgressView( postTaxSalaryString, value: postTaxSalary / salaryNum *
100, total: 100 )
Text( "Tax" )
ProgressView( incomeTaxString, value: incomeTax / salaryNum * 100,
total: 100 )
Text( "National Insurance" )
ProgressView( nationalInsuranceTaxSting, value: 13, total: 100 )
```

The reason for dividing the `postTaxSalary` and `incomeTax` variables by `salaryNum` and multiplying it by `100` is to work out the percentage for `ProgressView`. Before running, update the `VStack` to be returned. There seems to be an issue with the code, causing conflicts for the compiler. To resolve this, we need to explicitly state what to draw by returning it:

```
return VStack
```

If we run our application and insert the salary of £100,000, `ResultsView` will appear as follows:

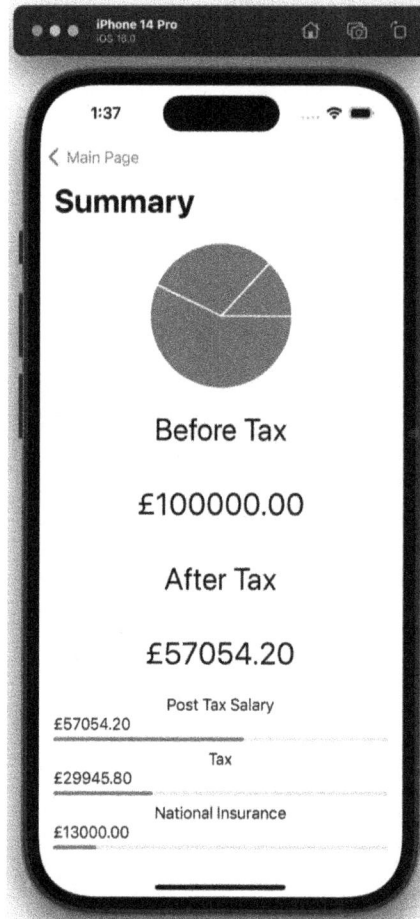

Figure 3.7 – ResultsView updated

That definitely was tiring, all that code. For reference, here is what the `ContentView` and `ResultsView` code should look like before moving on:

ContentView:

```
import SwiftUI

struct ContentView: View
{
    @State private var salary: String = ""
    @State private var isSalaryValid: Bool = false
```

```
var body: some View
{
    NavigationView
    {
        VStack
        {
            Text( "Annual Salary" )
                .padding(.bottom, 75.0)

            TextField( "", text: $salary )
                .frame( width: 200.0 )
                .border( Color.black, width: 1 )
                .padding( .bottom, 75.0 )
                .keyboardType( .decimalPad )

            NavigationLink( destination: ResultsView( salary:
$salary ), isActive: $isSalaryValid, label:
                {
                    Text( "Calculate Tax" )
                        .bold( )
                        .frame( width: 200, height: 50 )
                        .background( Color.blue )
                        .foregroundColor( Color.white )
                        .cornerRadius( 10 )
                        .onTapGesture
                        {
                            GoToResultsView( )
                        }
                } )
        }
        .padding( )
        .navigationTitle( "Main Page" )
    }
}

func GoToResultsView( )
{
    if ( nil != Float( salary ) )
    {
        if ( Float( salary )! > 0 )
        { isSalaryValid = true }
    }
}
```

```
}

struct ContentView_Previews: PreviewProvider
{
    static var previews: some View
    {
        ContentView( )
    }
}
```

ResultsView:

```
import SwiftUI
import SwiftUICharts

struct ResultsView: View
{
    @Binding var salary: String

    var body: some View
    {
        let salaryNum = Double( salary )!
        var incomeTax: Double = 0
        var nationalInsuranceTax: Double = 0

        if ( salaryNum > 12570 )
        {
            if ( salaryNum > 37700 )
            {
                if ( salaryNum > 150000 )
                {
                    incomeTax += ( 37700 - 12571 ) * 0.2
                    incomeTax += ( 150000 - 37701 ) * 0.4
                    incomeTax += ( salaryNum - 150000 ) * 0.45
                }
                else
                {
                    incomeTax += ( 37700 - 12571 ) * 0.2
                    incomeTax += ( salaryNum - 37700 ) * 0.4
                }
            }
            else
            { incomeTax += ( salaryNum - 12570 ) * 0.2 }
```

```
        }

        nationalInsuranceTax = salaryNum * 0.13

        let postTaxSalary = salaryNum - incomeTax -
nationalInsuranceTax

        let salaryString = String( format:"£%.2F", salaryNum )
        let postTaxSalaryString = String( format: "£%.2F",
postTaxSalary )
        let incomeTaxString = String( format: "£%.2F", incomeTax )
        let nationalInsuranceTaxSting = String( format: "£%.2F",
nationalInsuranceTax )

        let taxBreakdown: [Double] = [postTaxSalary, incomeTax,
nationalInsuranceTax]

        return VStack
        {
            PieChart( )
                .data( taxBreakdown )
                .chartStyle( ChartStyle( backgroundColor: .white,
                                          foregroundColor:
ColorGradient( .blue, .purple ) ) )

            Text( "Before Tax" )
                .font( .system( size: 32 ) )
                .padding(.vertical)

            Text( salaryString )
                .font( .system( size: 32 ) )
                .padding(.vertical)

            Text( "After Tax" )
                .font( .system( size: 32 ) )
                .padding(.vertical)

            Text( postTaxSalaryString )
                .font( .system( size: 32 ) )
                .padding(.vertical)

            Group
            {
                Text( "Post Tax Salary" )
```

```
                        ProgressView( postTaxSalaryString, value:
    postTaxSalary / salaryNum * 100, total: 100 )

                        Text( "Tax" )

                        ProgressView( incomeTaxString, value: incomeTax /
    salaryNum * 100, total: 100 )

                        Text( "National Insurance" )

                        ProgressView( nationalInsuranceTaxSting, value: 13,
    total: 100 )
                    }
                }
            .padding( )
            .navigationBarTitle( "Summary" )
        }
    }


struct ResultsView_Previews: PreviewProvider
{
    static var previews: some View
    {
        ResultsView( salary: .constant( "100" ) )
    }
}
```

In the next section, we will fix an error that occurs in ContentView; see whether you can figure out what the error is.

Fixing the ContentView binding error

There is an error in ContentView. It can be triggered by following these steps:

1. Navigate to ResultsView by clicking the **Calculate Tax** button.
2. Go back to ContentView by pressing the back arrow on the top left.
3. Input invalid data, which can be any of the following:

    A. Delete all the text in TextField

    B. Space character

    C. Any non-numerical character

    D. Two or more decimal points

Once these steps are complete, the following error will appear:

```
let salaryNum = Double( salary )!   ≡   Thread 1: Fatal error: Unexpectedly found nil while unwrapping an Optional value
```

Figure 3.8 – Invalid input error

This occurs because the variable is bound to `ResultsView`, which uses the value numerically. As the input is no longer numerical, it crashes. The problem is that `Double( salary )` returns `nil` and forces it to be assigned using `!`, which makes it crash. We will default assign a value of `0` to the `salaryNum` variable and if it's not `nil`, then cast the salary string as `Double`. Update the code as follows:

```
var salaryNum: Double = 0

if ( nil != Double( salary ) )
{
    salaryNum = Double( salary )!
}
```

Alternatively, you can use the coalescing operator to reduce the preceding check as follows: `let salaryNum = Double( salary ) ?? 0`. For more information on the coalescing operator, visit `https://docs.swift.org/swift-book/documentation/the-swift-programming-language/basicoperators/#Nil-Coalescing-Operator`.

First, we change the variable to `var` instead of `let` as it needs to be modifiable just for checking whether the cast equates to `nil`. Then, we check that it's not `nil` and assign the value accordingly. *Assigning 0 won't have any effect as the view isn't visible at this time.* Now if you run the application, it won't crash if you followed the preceding steps.

In the next section, we will rename `ContentView`.

Renaming ContentView to FrontView

In this small section, we will change `ContentView`'s name to `FrontView`. The default name of `ContentView` doesn't provide much information to us. We will rename it `FrontView`. You could rename the file and manually change every occurrence of `ContentView` to `FrontView`, which wouldn't be too tedious in an application this size. However, in a larger and more complex application, it would take away a lot of valuable development time. We can use Xcode's renaming tool to aid us.

Simply go to any reference to `ContentView` in the application's code, right-click it, and go to **Refactor | Rename…**:

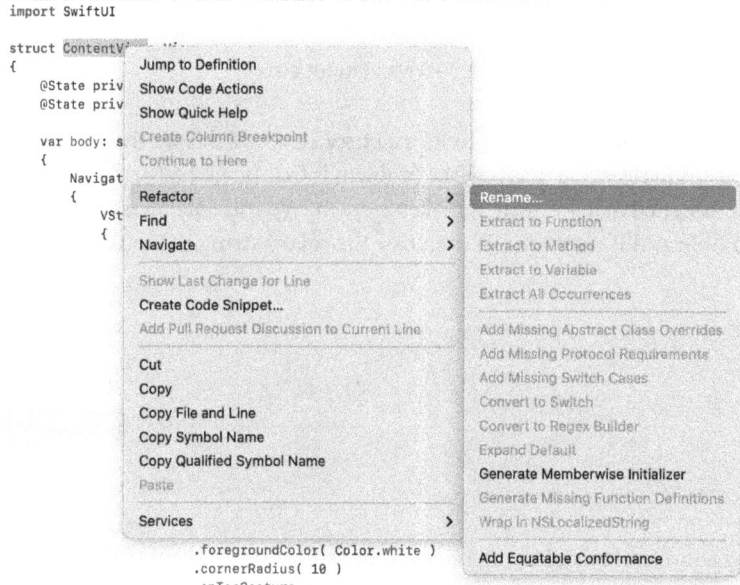

Figure 3.9 – Rename… button

In the next view, set the name to `FrontView` and press *Enter*:

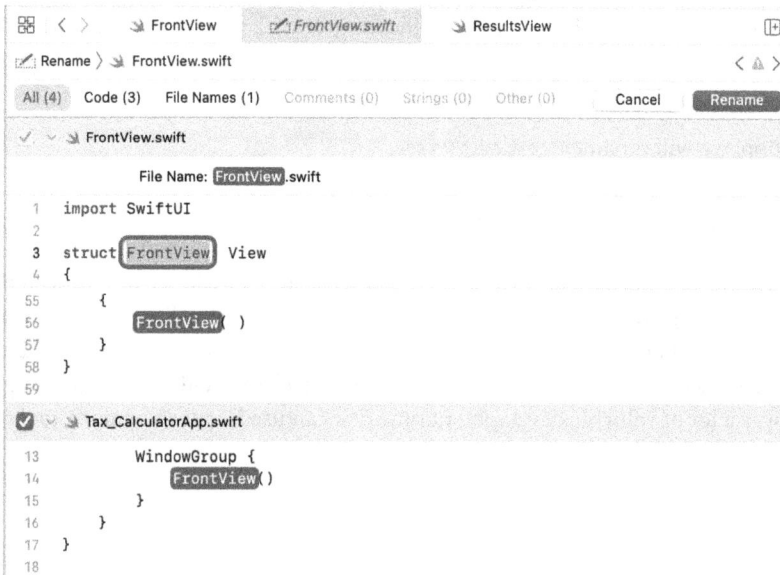

Figure 3.10 – Xcode renaming tool

Renaming files and updating all references is as easy as that. This also works for variables and functions; feel free to use it whenever you need to.

We have now completed the calculator application; in the next section, you will see some of the extra tasks for you to complete.

Extra tasks

Now that the application is complete, here is a list of tasks for you to complete to enhance your application and also to test your knowledge of the concepts learned in this chapter:

- Show an error when submitting an invalid salary
- Formatting salary and tax components with commas
- Abstracting tax brackets and percentages to make the app more dynamic
- Pie chart labeling
- Different types of tax
- Corporation tax
- Inheritance tax
- Provide more inputs for the current salary calculator such as pension contributions and student loans
- Back button styling

> **Note**
> If at any point you require help, feel free to join my Discord group at `https://discord.gg/7e78FxrgqH`.

We will summarize what we have covered in this chapter. However, before that, I will provide additional code for the extra tasks for you to implement at your own leisure.

Different tax options

To add different tax options to your code, you can modify `ResultsView` to include a picker or segmented control for selecting different tax options. Here's an example of how you can modify your code to add a picker for tax options:

```
import SwiftUI

struct FrontView: View {
```

```
    @State private var salary: String = ""
    @State private var isSalaryValid: Bool = false
    @State private var selectedTaxOption: TaxOption = .option1 //
Default tax option

    enum TaxOption: String, CaseIterable {
        case option1 = "Income Tax"
        case option2 = "Dividen Tax"
        case option3 = "Corporation Tax"

        // You can add more tax options if needed
    }

    var body: some View {
        NavigationView {
            VStack {
                Text("Annual Salary")
                    .padding(.bottom, 75.0)

                TextField("", text: $salary)
                    .frame(width: 200.0)
                    .border(Color.black, width: 1)
                    .padding(.bottom, 75.0)
                    .keyboardType(.decimalPad)

                Picker("Tax Option", selection: $selectedTaxOption) {
                    ForEach(TaxOption.allCases, id: \.self) { option
in
                        Text(option.rawValue)
                    }
                }
                .pickerStyle(SegmentedPickerStyle())
                .padding(.bottom, 75.0)

                NavigationLink(destination: ResultsView(salary:
$salary, taxOption: selectedTaxOption), isActive: $isSalaryValid) {
                    Text("Calculate Tax")
                        .bold()
                        .frame(width: 200, height: 50)
                        .background(Color.blue)
                        .foregroundColor(Color.white)
                        .cornerRadius(10)
                        .onTapGesture {
                            goToResultsView()
```

```
                            }
                        }
                    }
                    .padding()
                    .navigationTitle("Main Page")
                }
            }

    func goToResultsView() {
        if let salaryFloat = Float(salary), salaryFloat > 0 {
            isSalaryValid = true
        }
    }
}

struct ResultsView: View {
    var salary: String
    var taxOption: FrontView.TaxOption

    var body: some View {
        VStack {
            Text("Results")
                .font(.title)
                .padding()

            Text("Salary: \(salary)")
                .padding()

            Text("Tax Option: \(taxOption.rawValue)")
                .padding()

            // Calculate and display tax results based on the selected
tax option

            Spacer()
        }
        .navigationTitle("Results")
    }
}

struct ContentView_Previews: PreviewProvider {
    static var previews: some View {
        FrontView()
```

```
        }
    }
    ` ` `
```

In this example, I've added the `TaxOption` enum to define different tax options. The `selectedTaxOption` property is used to store the selected tax option. I've added a picker (using a segmented control style) to allow the user to select the tax option.

When the user taps **Calculate Tax**, the selected tax option is passed to `ResultsView`, where you can calculate and display the tax results based on the selected option.

Tax geography

To add the ability to calculate tax breakdowns for different geographies, including countries and states, you can modify `ResultsView` and introduce a selection mechanism for geographies. Here's an example of how you can modify your code to achieve that:

```
import SwiftUI

struct FrontView: View {
    @State private var salary: String = ""
    @State private var isSalaryValid: Bool = false
    @State private var selectedGeography: Geography = .country("USA")
// Default geography

    enum Geography {
        case country(String)
        case state(String, String) // Country, State

        var displayText: String {
            switch self {
            case .country(let country):
                return country
            case .state(let country, let state):
                return "\(state), \(country)"
            }
        }
    }

    var body: some View {
        NavigationView {
            VStack {
                Text("Annual Salary")
                    .padding(.bottom, 75.0)
```

```
                TextField("", text: $salary)
                    .frame(width: 200.0)
                    .border(Color.black, width: 1)
                    .padding(.bottom, 75.0)
                    .keyboardType(.decimalPad)

                Picker("Geography", selection: $selectedGeography) {
                    Text("USA").tag(Geography.country("USA"))
                    Text("California, USA").tag(Geography.state("USA",
"California"))
                    // Add more geography options as needed
                }
                .pickerStyle(SegmentedPickerStyle())
                .padding(.bottom, 75.0)

                NavigationLink(destination: ResultsView(salary:
$salary, geography: selectedGeography), isActive: $isSalaryValid) {
                    Text("Calculate Tax")
                        .bold()
                        .frame(width: 200, height: 50)
                        .background(Color.blue)
                        .foregroundColor(Color.white)
                        .cornerRadius(10)
                        .onTapGesture {
                            goToResultsView()
                        }
                }
            }
            .padding()
            .navigationTitle("Main Page")
        }
    }

    func goToResultsView() {
        if let salaryFloat = Float(salary), salaryFloat > 0 {
            isSalaryValid = true
        }
    }
}

struct ResultsView: View {
    var salary: String
    var geography: FrontView.Geography
```

```
    var body: some View {
        VStack {
            Text("Results")
                .font(.title)
                .padding()

            Text("Salary: \(salary)")
                .padding()

            Text("Geography: \(geography.displayText)")
                .padding()

            // Calculate and display tax breakdown based on the
selected geography

            Spacer()
        }
        .navigationTitle("Results")
    }
}

struct ContentView_Previews: PreviewProvider {
    static var previews: some View {
        FrontView()
    }
}
```

In this example, I've introduced the Geography enum, which represents different geographies. It can be either a country or a state within a country. The selectedGeography property is used to store the selected geography.

In FrontView, I've added a picker (using a segmented control style) to allow the user to select the geography. You can add more geography options as needed.

When the user taps **Calculate Tax**, the selected salary and geography are passed to ResultsView, where you can calculate and display the tax breakdown based on the selected geography.

In ResultsView, I've added a displayText computed property to format and display the selected geography. You can modify the code to calculate and display the tax breakdown based on the selected geography.

> **Note**
>
> This code provides a basic structure for adding tax breakdowns based on geographies. The actual tax calculations and breakdowns would need to be implemented based on the tax laws and rules of the specific countries and states involved.

Here's the updated code for `ResultsView` with the tax breakdown based on the given logic and the addition of the `Geography` selection:

ResultsView

```swift
import SwiftUI
import SwiftUICharts

struct ResultsView: View {
    @Binding var salary: String
    var geography: FrontView.Geography

    var body: some View {
        let salaryNum: Double = Double(salary) ?? 0
        let taxBreakdown = calculateTaxBreakdown(for: salaryNum, in:
geography)

        let salaryString = formatCurrency(salaryNum)
        let postTaxSalaryString = formatCurrency(taxBreakdown.
postTaxSalary)
        let incomeTaxString = formatCurrency(taxBreakdown.incomeTax)
        let nationalInsuranceTaxString = formatCurrency(taxBreakdown.
nationalInsuranceTax)

        return VStack {
            PieChart()
                .data(taxBreakdown.chartData)
                .chartStyle(ChartStyle(backgroundColor: .white,
foregroundColor: ColorGradient(.blue, .purple)))

            Text("Before Tax")
                .font(.system(size: 32))
                .padding(.vertical)

            Text(salaryString)
                .font(.system(size: 32))
                .padding(.vertical)
```

```
                Text("After Tax")
                    .font(.system(size: 32))
                    .padding(.vertical)

                Text(postTaxSalaryString)
                    .font(.system(size: 32))
                    .padding(.vertical)

                Group {
                    Text("Post Tax Salary")

                    ProgressView(postTaxSalaryString, value: taxBreakdown.
postTaxPercentage, total: 100)

                    Text("Tax")

                    ProgressView(incomeTaxString, value: taxBreakdown.
incomeTaxPercentage, total: 100)

                    Text("National Insurance")

                    ProgressView(nationalInsuranceTaxString, value:
taxBreakdown.nationalInsurancePercentage, total: 100)
                }
            }
            .padding()
            .navigationBarTitle("Summary")
    }

    private func calculateTaxBreakdown(for salary: Double, in
geography: FrontView.Geography) -> TaxBreakdown {
        var incomeTax: Double = 0
        var nationalInsuranceTax: Double = 0

        if salary > 12570 {
            if salary > 37700 {
                if salary > 150000 {
                    incomeTax += (37700 - 12571) * 0.2
                    incomeTax += (150000 - 37701) * 0.4
                    incomeTax += (salary - 150000) * 0.45
                } else {
                    incomeTax += (37700 - 12571) * 0.2
                    incomeTax += (salary - 37700) * 0.4
                }
```

```
            } else {
                incomeTax += (salary - 12570) * 0.2
            }
        }

        nationalInsuranceTax = salary * 0.13

        let postTaxSalary = salary - incomeTax - nationalInsuranceTax
        let chartData: [(String, Double)] = [("Post Tax Salary",
postTaxSalary), ("Tax", incomeTax), ("National Insurance",
nationalInsuranceTax)]

        let totalSalary = salary + nationalInsuranceTax
        let postTaxPercentage = postTaxSalary / totalSalary * 100
        let incomeTaxPercentage = incomeTax / totalSalary * 100
        let nationalInsurancePercentage = nationalInsuranceTax /
totalSalary * 100

        return TaxBreakdown(postTaxSalary: postTaxSalary, incomeTax:
incomeTax, nationalInsuranceTax: nationalInsuranceTax, chartData:
chartData, postTaxPercentage: postTax

Percentage, incomeTaxPercentage: incomeTaxPercentage,
nationalInsurancePercentage: nationalInsurancePercentage)
    }

    private func formatCurrency(_ value: Double) -> String {
        let formatter = NumberFormatter()
        formatter.numberStyle = .currency
        formatter.currencySymbol = "£"

        return formatter.string(from: NSNumber(value: value)) ?? ""
    }
}

struct TaxBreakdown {
    var postTaxSalary: Double
    var incomeTax: Double
    var nationalInsuranceTax: Double
    var chartData: [(String, Double)]
    var postTaxPercentage: Double
    var incomeTaxPercentage: Double
    var nationalInsurancePercentage: Double
}
```

```
struct ResultsView_Previews: PreviewProvider {
    static var previews: some View {
        ResultsView(salary: .constant("100"), geography:
.country("USA"))
    }
}
```

In this updated code, `ResultsView` now accepts a `geography` parameter of type `FrontView`. `Geography`. The `calculateTaxBreakdown(for:in:)` function is added to calculate the tax breakdown based on the salary and selected geography.

The tax breakdown is stored in the `TaxBreakdown` struct, which includes the post-tax salary, income tax, national insurance tax, chart data, and percentage values for each tax category.

The `formatCurrency(_:)` function is used to format currency values with the pound sign (£).

The calculated tax breakdown is used to populate the `PieChart` and `ProgressView` components to display the tax breakdown and percentages.

Please note that the tax calculation logic in the code provided is based on the initial logic and may not be accurate or applicable to real-world tax calculations. It serves as an example structure for integrating tax breakdown calculations into your SwiftUI code.

Summary

In this chapter, we implemented all the calculator's functionality. We linked all the UI components that we implemented in the previous chapter. First, we provided a means to navigate to and from `ResultsView`. Then, we checked the salary input to make sure it was above zero and didn't contain any invalid characters. Once validated, we passed the salary from `ContentView` to `ResultsView`. Using the salary, we calculated the tax breakdown in `ResultsView`, fixed an annoying error, and renamed `ContentView` to `FrontView`. Finally, we also implemented a few extra tasks for our tax calculator app.

In the next chapter, we'll start our next application, which will be a photo gallery for the iPad. We will leverage many of the skills learned already, so feel free to take a moment and go back over anything you didn't fully understand.

4
iPad Project – Photo Gallery Overview

In the previous two chapters, we created a Tax Calculator app for the iPhone. We implemented it from scratch, and we looked at the technical requirements, design specifications, wireframes, and code implementation. We will use these newly acquired skills in this and the next chapter, but worry not, we will go over all necessary aspects in case you have jumped straight to this chapter.

In this chapter, we will work on the design of our second project, a photo gallery application for the iPad that will showcase its big screen. We will assess the requirements for designing such an application and discuss the design specifications, allowing us to get a better understanding of what is required and how it will all fit together. Then we will start coding our application to build out the UI, which will be connected to the gallery's enhanced view in the next chapter. This project will cover the foundations of SwiftUI components. We will discuss all this in the following sections:

- Requirements
- Design Specifications
- Building the Gallery UI

By the end of this chapter, you will have a better understanding of what features/components are required in our photo gallery application, how we will utilize the larger screen of the iPad (compared to the iPhone), and the design of our application. By the end of this chapter, we will have implemented the wireframe for the first page, which will serve as the foundation for the next chapter.

Technical Requirements

This chapter requires you to download Xcode version 14 or above from Apple's App Store.

To install Xcode, just search for Xcode in the App Store and download the latest version. Open Xcode and follow the installation instructions. Once Xcode has opened and launched, you're ready to go.

Version 14 of Xcode has the following features/requirements:

- Includes SDKs for iOS 16, iPadOS 16, macOS 12.3, tvOS 16, and watchOS 9.

- Supports on-device debugging in iOS 11 or later, tvOS 11 or later, and watchOS 4 or later.

- Requires a Mac running macOS Monterey 12.5 or later.

- For further information regarding technical details, please refer to *Chapter 1*.

The code files for this chapter can be found here: `https://github.com/PacktPublishing/Elevate-SwiftUI-Skills-by-Building-Projects`.

In the next section, we will provide clarity on the specifications of our application's design and look at mockups of what the app will look like.

Understanding the Design Specifications

In this section, we will look at the design specifications of our gallery application. This section describes the features we are going to implement in our gallery app. The best method for figuring out the features required is to put yourself in the user's shoes to determine how they will use the app and break it into individual steps.

The features of our app we would like to present are as follows:

- **Highlight view**: This is the main view that the user will see, which showcases all the images.

- **Enhanced view**: This shows a larger version of the image along with information such as a description and a date.

- **Fullscreen mode**: View the image in fullscreen mode without any extra information.

- **Fullscreen tap for more info**: A single tap while in fullscreen mode will show the photo's title.

- **Collections**: Different image collections, or albums:

  - **Side panel**: This shows all the collection names in horizontal mode.

- **Delete and rename**: This allows the user to delete an image and rename it.

- **Editing**: Image editing functionality:

  - Drawing using the Apple Pencil.

- **Sharing**: The ability to share an image.

- **Display mode switching**: Change how the images are displayed on the Highlight page, as a list or using tiles.

- **Annotations and highlighting**: Allow the user to annotate and highlight parts of the image.

- Images and information come from an external source such as a local database or online.

Now that we have listed the ideal features we would like, it is important for us to determine which features are crucial. To do this, we must understand the purpose of our product. For me, the purpose of creating this photo gallery app is to showcase some images and provide an enhanced mode that gives more information. I know that not all the features are required, and it could be useful if some were omitted and assigned as extra tasks for you as the developer to undertake. The extra tasks are optional and should only be performed when you've finished this project. With this in mind, the following are the core features we will be implementing:

- Highlight view

- Enhanced view

- Sharing

The rest of the features will be an exercise for you once you have completed this and the next chapter. The next section will cover the acceptance criteria for our application.

Acceptance criteria

We will discuss the acceptance criteria for our application that we want to see in the end product, which will be completed at the end of the next chapter. If possible, we should try to make them measurable. Let's do this right now:

- Utilize small images for the gallery view, providing an overview of the whole gallery.

- A scrollable view to navigate through all our images.

- An enhanced view to display extra information, thus allowing optional parameters.

- A native sharing menu should appear when sharing an image.

- Navigation from the highlight view to the enhanced view.

- Navigation from the enhanced view to the fullscreen view.

Develop test cases in which the application's acceptance criteria will be tested. Using this method allows us to see the conditions under which the application will be used by the end user and the level that needs to be attained for it to be considered successful.

Wireframe

One of the most useful tools for designing layouts is wireframing. A wireframe is an overview of the layout. The following image shows what the highlight page of our app will look like using a wireframe:

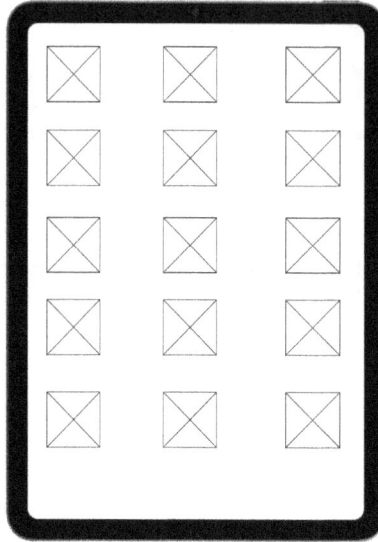

Figure 4.1 – Wireframe of the Highlight view in portrait mode

With iPad applications, it is very important to have a landscape mode. The following image shows the wireframe of our highlight view in landscape mode:

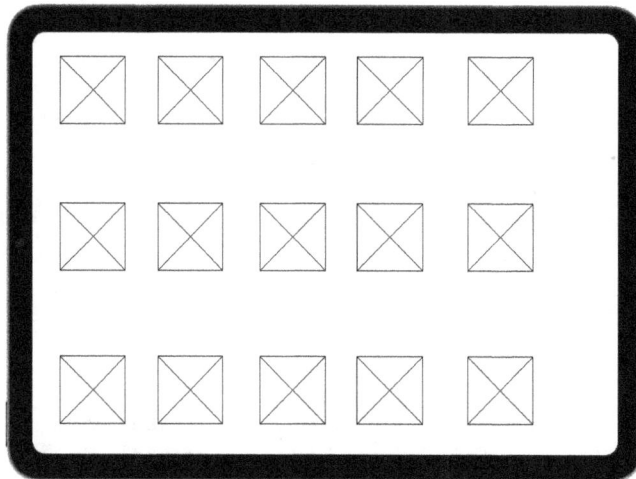

Figure 4.2 – Wireframe of the Highlight view in landscape mode

The following two images show the portrait and landscape wireframes for the enhanced view:

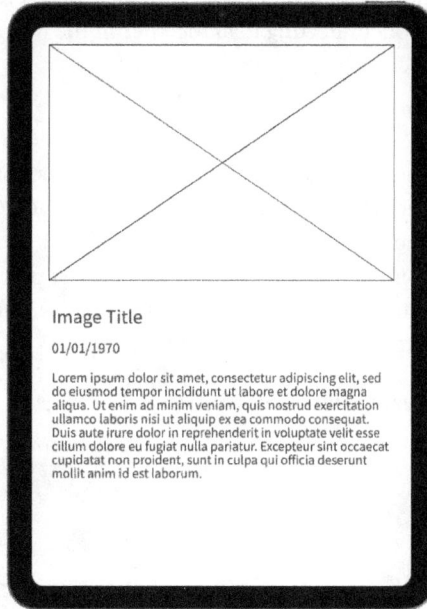

Figure 4.3 – Wireframe of the Enhanced view in portrait mode

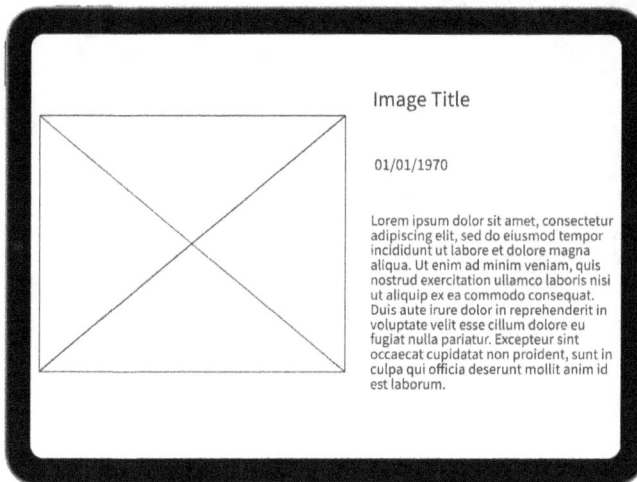

Figure 4.4 – Wireframe of the Enhanced view in landscape mode

In the next section, we will build the interface for our application and make sure it looks the way we designed it in the wireframes. Though we will build it the same way, there may be small differences. This will serve as the foundation for connecting it all together in the next chapter.

Building the Gallery UI

We will now build the UI for the gallery app. There are three main parts of the gallery, the first being the highlight page, which is loaded on launch and shows all the images. Once the user clicks on an image, the user is taken to the enhanced page, which is the second part. On this page, a bigger version of the image is shown along with more information. Finally, the last part is fullscreen mode, which simply shows a selected image in fullscreen. Naturally, we will start with the first part, the highlight page, but before that, we will create our project. Follow these steps:

1. Open Xcode and select **Create a new Xcode project**:

Figure 4.5 – Create a new Xcode project

2. Now, we will choose the template for our application. As we are creating an iPad application, we will select **iOS** from the top, then select **App**, and click **Next**:

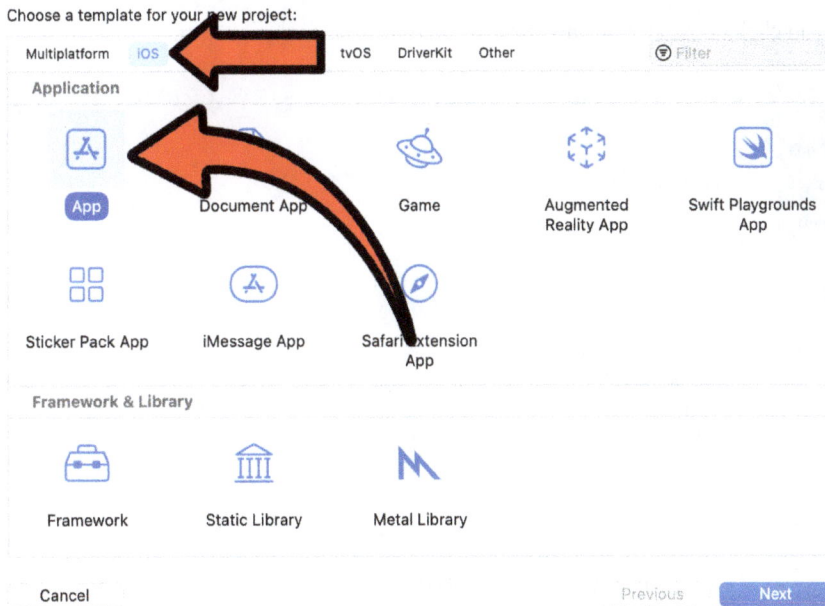

Figure 4.6 – Xcode project template selection

3. We will now choose the options for our project. There are only two crucial things to set. Make sure **Interface** is set to **SwiftUI**. This will be the UI our system will use. Set **Language** to **Swift**; this is the programming language used for our application:

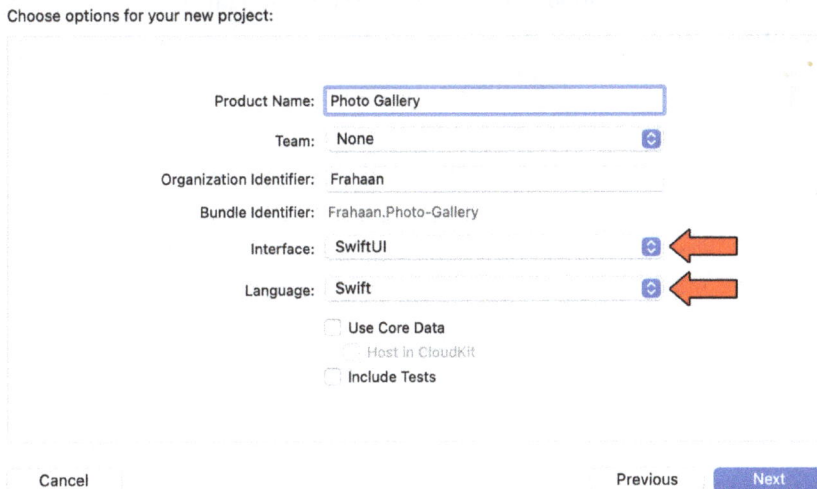

Figure 4.7 – Xcode project options

4.   Once you click on **Next**, you can choose where to create your project:

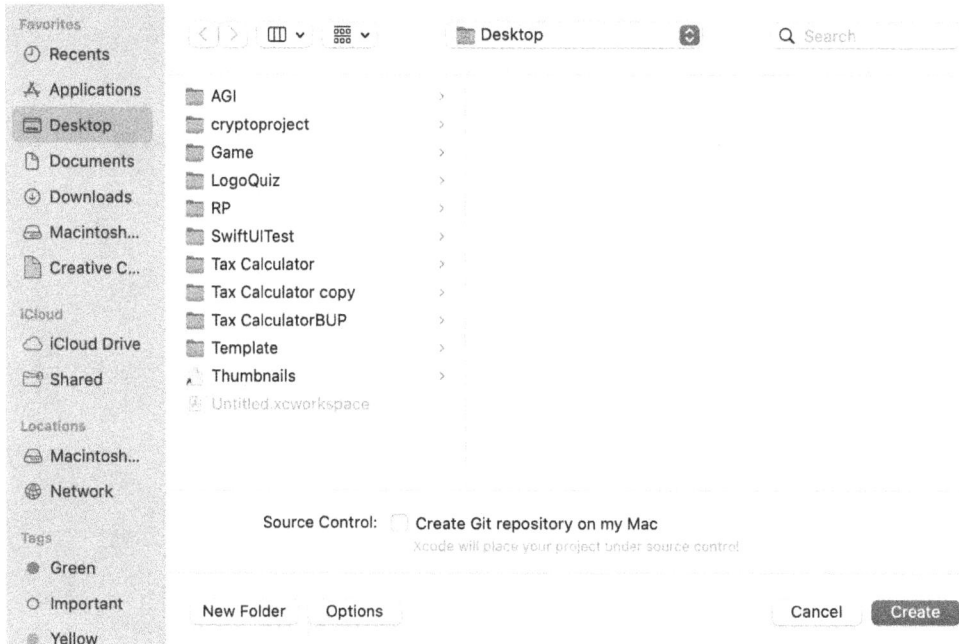

Figure 4.8 – Xcode project save directory

5.   Once you have chosen a location, click **Create** in the bottom right. Xcode shows your project in all its glory:

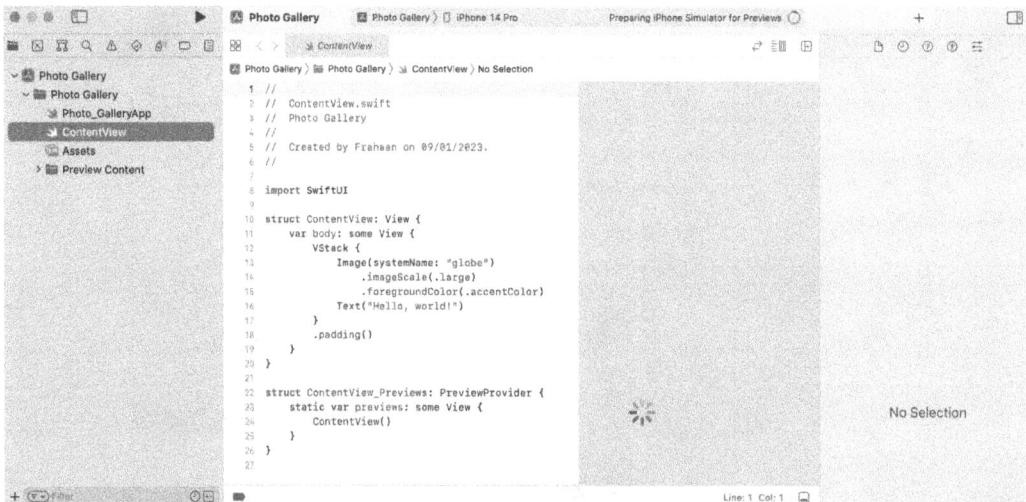

Figure 4.9 – New Xcode project overview

In the next section, we will set our project to iPad only.

Set Project to iPad

You have probably noticed in the preview that it's currently set to an iPhone project or, more accurately, an iOS project. This means it runs on iPhones and iPads, and the preview by default shows the iPhone view. We want our project to be iPad only, so follow these steps to set our project to iPad only:

1. Let's change it now. Select the project in the **Project Navigator**:

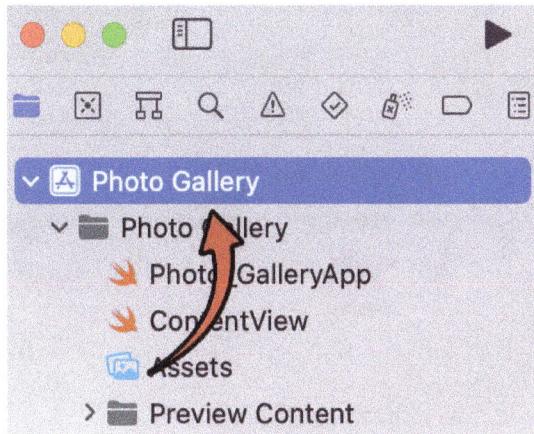

Figure 4.10 – Project in Project Navigator

2. Now select **Photo Gallery** from the **TARGETS** section:

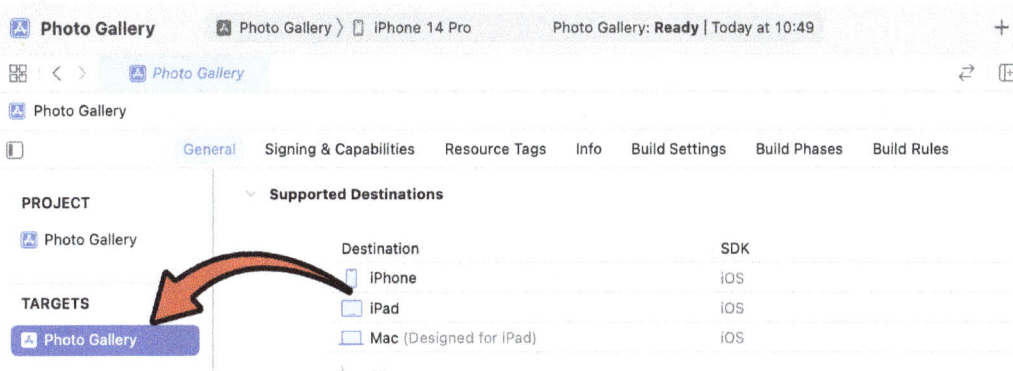

Figure 4.11 – Target selection

3. The next step is to remove the iPhone and Mac destinations. Doing this is simple; select each destination and press the minus button:

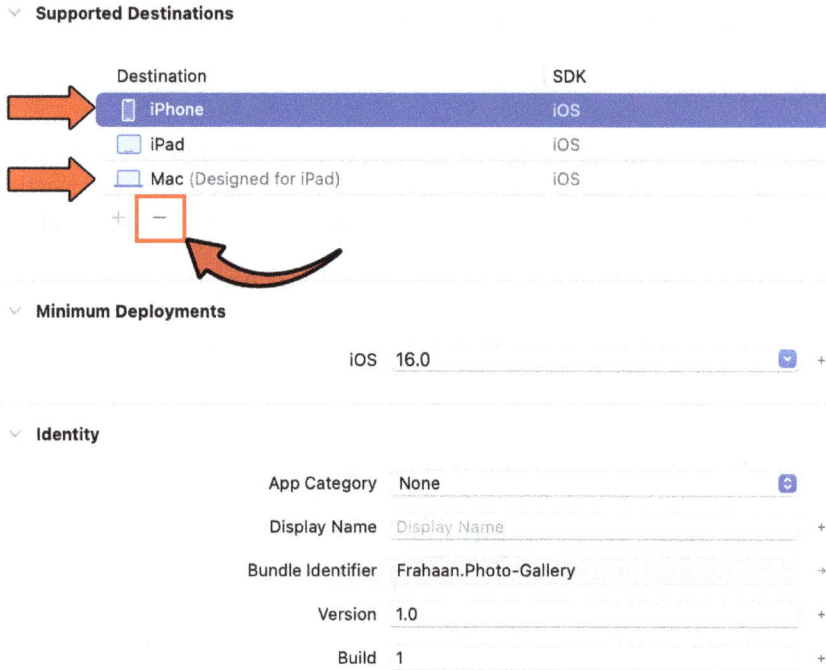

Figure 4.12 – Remove destinations

4.   After removing iPhone and Mac from the destinations, it should look as follows:

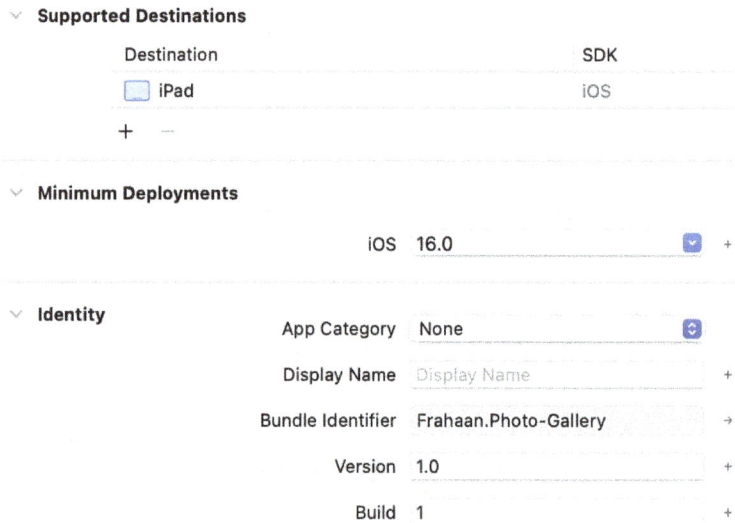

Figure 4.13 – Supported destinations

In the next section, we will implement the highlight page of our application using SwiftUI.

Highlight Page

In this section, we will implement the highlight page's UI. As a reminder, this is what it will look like:

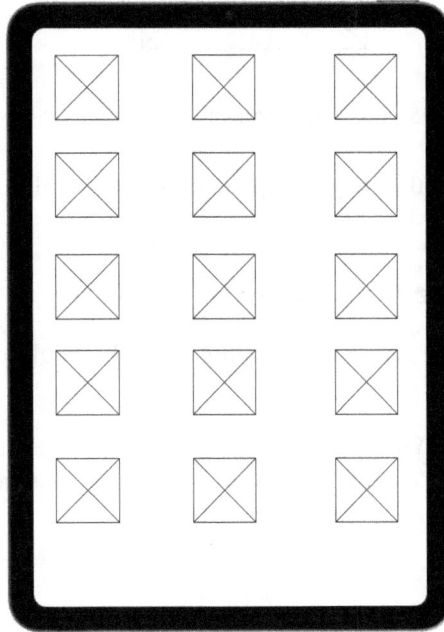

Figure 4.14 – Wireframe of the Highlight view in portrait mode

There is one main element on the highlight page. As a little task, see if you can figure out what it is. Don't worry if you don't know the exact UI component name; we will look at it in the following section.

Image Component

The image component is one of the core components provided by SwiftUI. It allows you to display an image, which can be used to provide a visual representation of something or to embellish a body of text. We will use it to show all the images in our gallery as highlights on our Highlight page. The following image shows the image on the Highlight page. As you will remember from *Figure 4.14*, The Highlight page is full of small images. These images will be clickable, which will take the user to the Enhanced view.

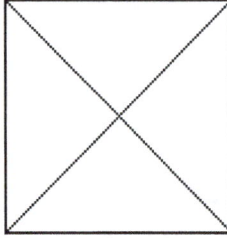

Figure 4.15 – Image component on the Highlight view

In the next section, we will add the image components we discussed earlier into our application using SwiftUI.

Adding Highlight Page Components

In this section, we will add the image components to our highlight page, which currently is named **ContentView**:

1. First, let's rename **ContentView** `HighlightView`. Doing this is simple: open **ContentView**, right-click any reference **ContentView** in the code, then go to **Refactor | Rename**. I will use the first reference, as shown in the following screenshot:

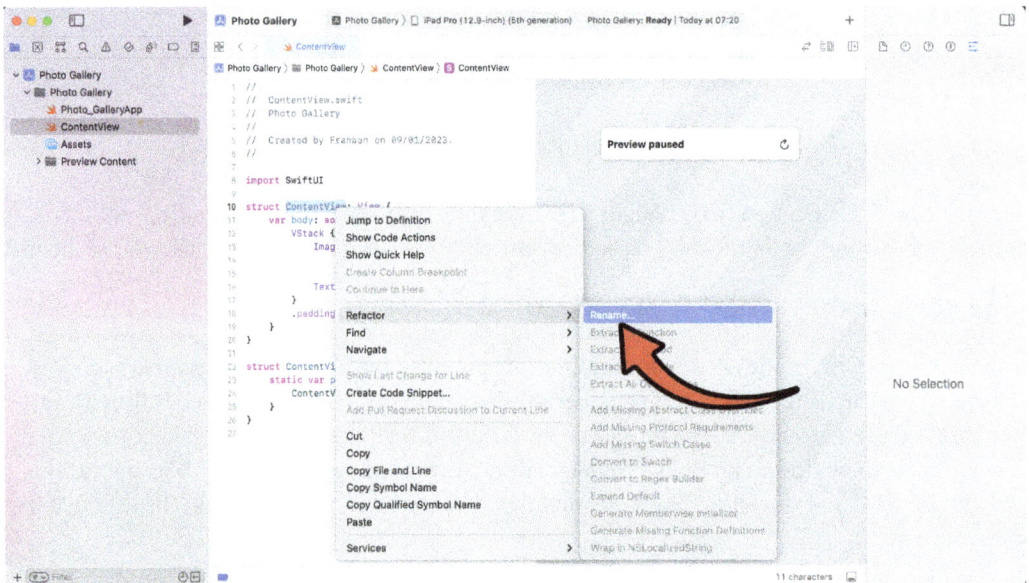

Figure 4.16 – Rename button

2. Next, the following screen will be shown:

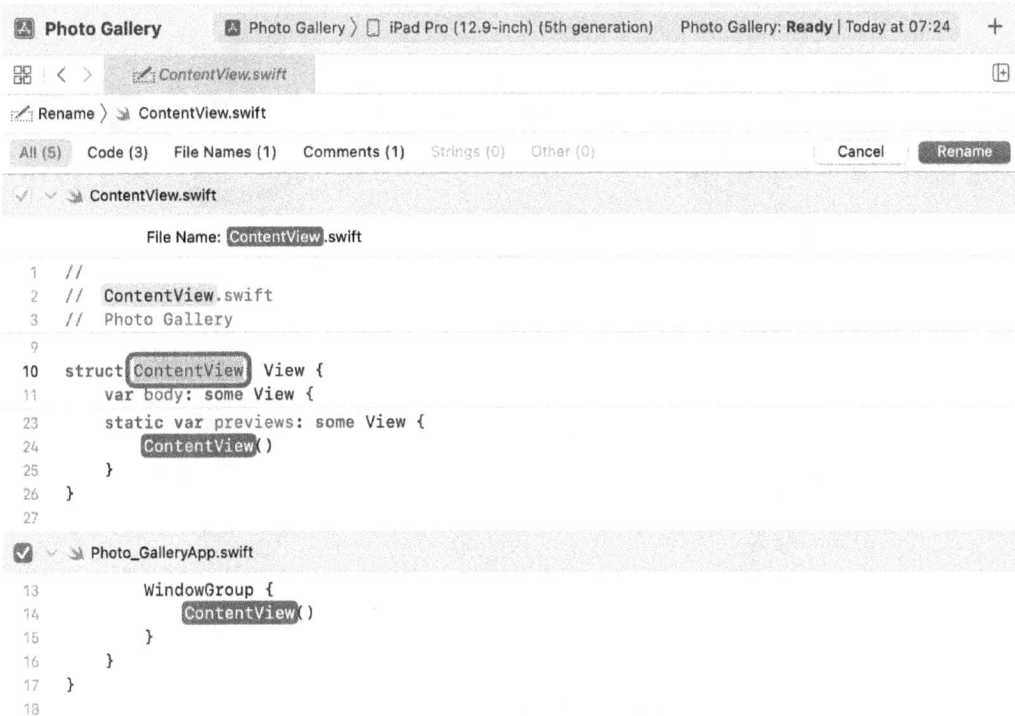

Figure 4.17 – Rename screen

3. Change the name from **ContentView** to HighlightView, and as you can see, all other references to **ContentView** are automatically updated. Finally, click the **Rename** button in the top right:

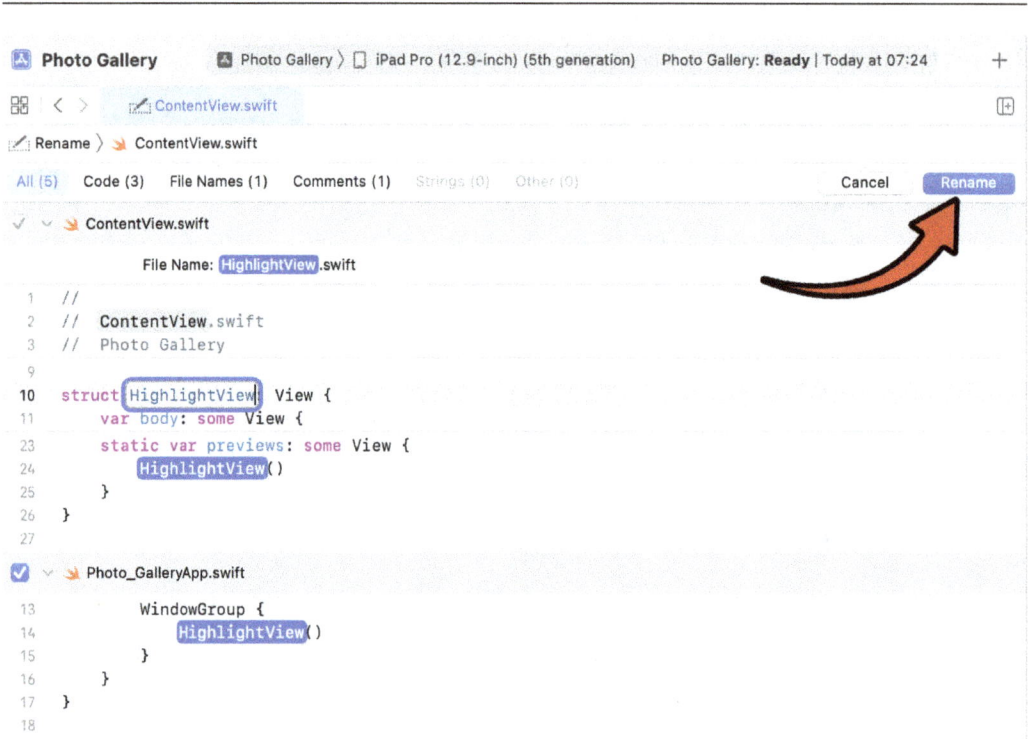

Figure 4.18 – Changing the view's name

4.   We have now renamed our view, including the file, as can be seen in the **Project Navigator**:

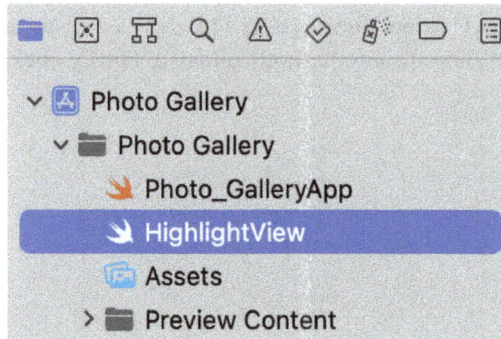

Figure 4.19 – Updated filename in the Project Navigator

5. There is one extra step, which is optional. That is to rename the `ContentView_Previews` struct. Though this is not crucial, I highly recommend renaming it to keep all the name references in sync. Using the same process as before, rename the `ContentView_Previews` struct `HighlightView_Previews`. The location of the struct is at the bottom of the `HighlightView` file (previously named `ContentView`):

```swift
 8  import SwiftUI
 9
10  struct HighlightView: View {
11      var body: some View {
12          VStack {
13              Image(systemName: "globe")
14                  .imageScale(.large)
15                  .foregroundColor(.accentColor)
16              Text("Hello, world!")
17          }
18          .padding()
19      }
20  }
21
22  struct ContentView_Previews: PreviewProvider {
23      static var previews: some View {
24          HighlightView()
25      }
26  }
27
```

Figure 4.20 – Renaming the previews struct

In the next section, we will take a look at column layouts and determine which one will be best for our application.

Column layouts

Before we start coding our highlight page, let's discuss the three main types of column layouts:

- **Flexible layout**: Allows you to specify the number of columns and spaces them according to the screen size dynamically.

- **Fixed layout**: Creates columns with fixed dimensions, which is restrictive.

- **Adaptive columns**: As the name suggests, they adapt to the size of the contents. You set the minimum size, which is used by the adaptive system to calculate how many items can be placed on a single row, depending on the screen size. Naturally, more items will be displayed when the iPad is in horizontal mode than in portrait mode.

Each has its use cases, but to truly show off our photo gallery, we will use the adaptive column system. In the next section, we will implement the highlight view programmatically.

Implementing the Highlight View

As we have created a fresh project, the coding standards aren't in line with my personal preferences. So, firstly, I will change the standards. Feel free to take a few moments to do the same:

1. First, we need to add the images to our project. Doing this is simple. Select **Assets** from the **Project Navigator**:

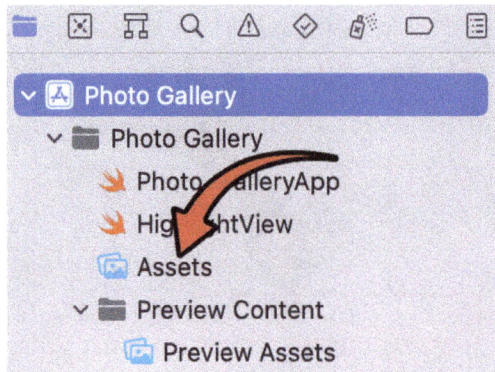

Figure 4.21 – Assets location in Project Navigator

2. Now the **Assets** view will appear. Importing an image/asset can be done in one of two ways:

   I. Dragging and dropping the files into the **Assets** section:

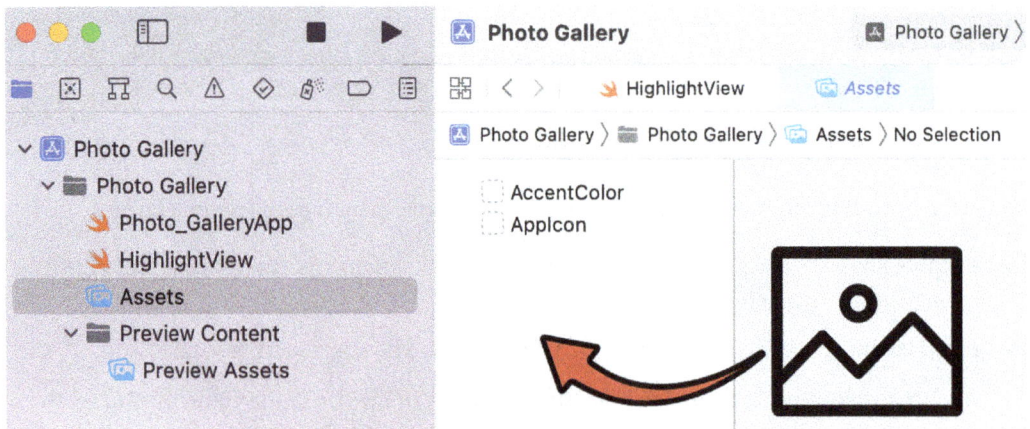

Figure 4.22 – Dragging and dropping Assets

II.   Right-clicking the **Assets** section and selecting **Import**:

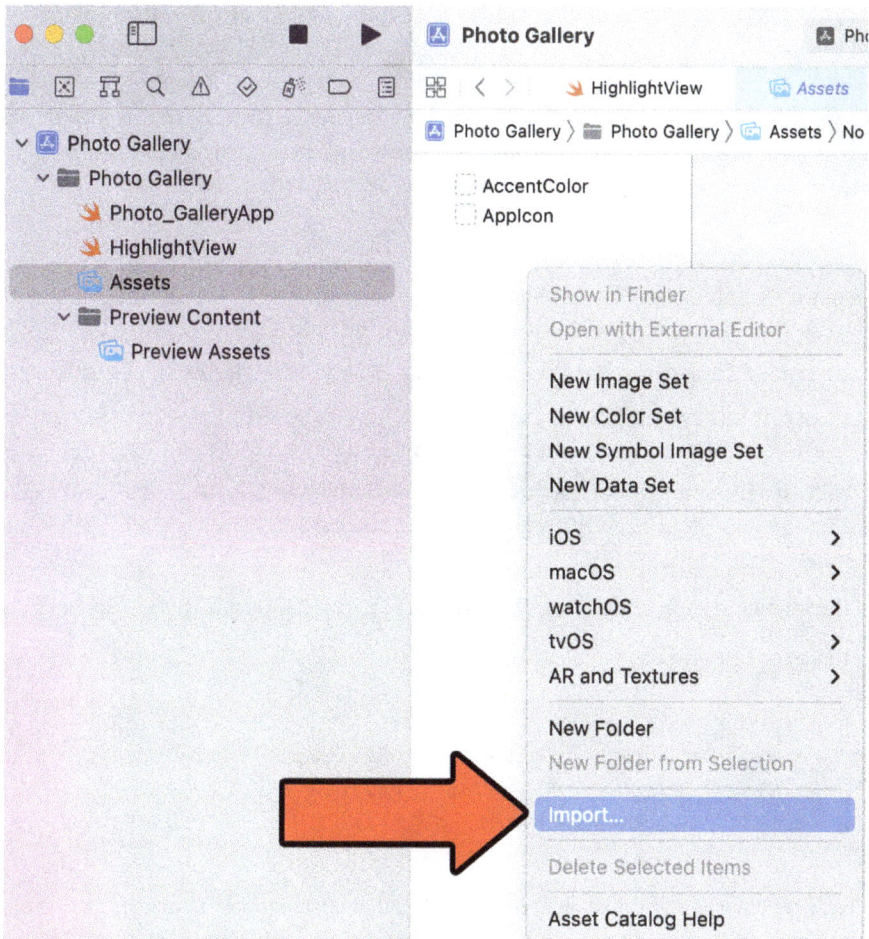

Figure 4.23 – Import button

3. Once the asset(s) have been imported, the **Assets** view will look as follows:

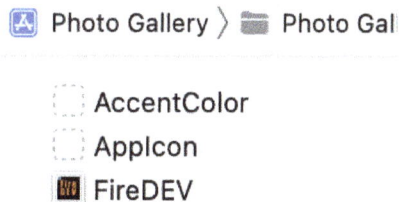

Figure 4.24 – Asset(s) imported

> **Tip**
>
> I have only imported one image, and I will use it multiple times, but you can and should use different images.

Next, we will create an array of strings. Each string will be the name of an image in our gallery. As mentioned previously, I will be using the same image, hence why all the strings will be the same. Update it according to your requirements.

> **Note**
>
> I am using the thumbnail for my developer-centric podcast FireDEV. Feel free to use it and tune in every Thursday to my podcast using the following links:
>
> Spotify: `https://open.spotify.com/show/387RiHksQE33KYHTitFXhg`
>
> Apple Podcasts: `https://podcasts.apple.com/us/podcast/firedev-fireside-chat-with-industry-professionals/id1602599831`
>
> Google Podcasts: `https://podcasts.google.com/feed/aHR0cHM6Ly9hbmNob3IuZm0vcy83Yjg2YTNiNC9wb2RjYXN0L3Jzc.`

4. Add the following code to the start of the `HighlightView` struct, before the body:

```
private let images: [String] =
[
    "FireDEV", "FireDEV", "FireDEV", "FireDEV", "FireDEV",
"FireDEV", "FireDEV", "FireDEV", "FireDEV", "FireDEV",
"FireDEV", "FireDEV", "FireDEV", "FireDEV", "FireDEV",
"FireDEV", "FireDEV", "FireDEV", "FireDEV", "FireDEV"
]
```

The code we have added is just an array of strings that correspond to image files. The file type, for example, `.png`, doesn't need to be included when specifying assets in Swift.

5. Next, we will create a variable, which sets how the columns will be organized. Earlier, we discussed the different column layouts and chose adaptive columns. Add the following code underneath the previous code:

```
private let adaptiveColumns =
[
    GridItem( .adaptive( minimum: 300 ) )
]
```

This code sets the grid items to be adaptive, with a minimum size of 300 pixels, which means regardless of the iPad size, the images will be easy to view and not too small or large.

6. Now, we are going to display the images. Add the following code inside the body, and we will go over what each part does:

```
ScrollView
{
    LazyVGrid( columns: adaptiveColumns, spacing: 20 )
    {
        ForEach( images.indices )
        { i in
            Image( images[i] )
                .resizable( )
                .scaledToFill( )
                .frame( width: 300, height: 300 )
        }
    }
}
```

Before we run it, let's see what each part of the code does:

- `LazyVGrid( columns: adaptiveColumns, spacing: 20 )`: Creates a lazy vertical grid with adaptive columns that have a spacing of 20 between each item.

- `ForEach( images.indices )`

    `{ i in`: Loops through every image filename, and i represents the index number, starting at 0, which will be used to reference the image shortly.

- `Image( images[i] )`

    `.resizable( )`

    `.scaledToFill( )`

    `.frame( width: 300, height: 300 )`: Creates an image using the filename from the `images` array. `resizable` allows an image's size to be modified. It's important to set the `resizable()` modifier before applying any scaling or sizing to an image, or it won't have an effect. `scaledToFill` is used to maintain the aspect ratio of an image and scales to fill its parent. Finally, the `frame` parameter is used to set the width and height of the image.

7.   Running the application will result in the following:

Figure 4.25 – HighlightView Preview

8.   Now let's rotate our device/simulator to see how it looks:

Figure 4.26 – Horizontal orientation preview

9.   Rotating the simulator can be achieved by pressing the ⌘ + *left/right arrow* keys or by pressing the rotate button at the top of the simulator, as shown in the following screenshot:

Figure 4.27 – Rotate button

As we can see, the Highlight view for our gallery application is looking really good. It works flawlessly in portrait and landscape. The beauty of using a grid instead of a custom implementation is that it handles the rotation, which saves time and energy, while keeping it in line with the design standards users are used to.

That was the last bit of code for this chapter. Here is all the code for the `HighlightView` file:

```swift
import SwiftUI

struct HighlightView: View
{
    private let images: [String] =
    [
        "FireDEV", "FireDEV", "FireDEV", "FireDEV", "FireDEV",
"FireDEV", "FireDEV", "FireDEV", "FireDEV", "FireDEV", "FireDEV",
"FireDEV", "FireDEV", "FireDEV", "FireDEV", "FireDEV", "FireDEV",
"FireDEV", "FireDEV", "FireDEV"
    ]

    private let adaptiveColumns =
    [
        GridItem( .adaptive( minimum: 300 ) )
    ]

    var body: some View
    {
        ScrollView
        {
            LazyVGrid( columns: adaptiveColumns, spacing: 20 )
            {
                ForEach( images.indices )
                { i in
                    Image( images[i] )
                        .resizable( )
                        .scaledToFill( )
                        .frame( width: 300, height: 300 )
```

```
                    }
                }
            }
        }
    }

    struct ContentView_Previews: PreviewProvider
    {
        static var previews: some View
        {
            HighlightView( )
        }
    }
```

We first looked at the design of our photo gallery application. It was unique from a design perspective due to it being on an iPad. We looked at the wireframe of the Highlight view, enhanced view, and full-screen mode. We used SwiftUI to implement the Highlight view and made sure it rotated correctly.

Summary

In this chapter, we covered the design of our photo gallery application for the iPad. We looked at wireframes and broke down each element into SwiftUI components. We then implemented the SwiftUI components to match the design from the highlight view wireframe. We also took a look at the requirements for building this application, and the design specifications a photo gallery application can have. Then we simplified it to the core features our app will provide. We expanded our design specifications with acceptance criteria to show what we would like our app to do.

In our next chapter, we'll take a look at implementing the enhanced view based on the wireframe discussed in this chapter and connect it to the highlight view.

iPad Project – Photo Gallery Enhanced View

In this chapter, we will work on implementing enhanced view and page navigation functionalities in our photo gallery project. In the previous chapter, we looked at the design of the photo gallery and how it works uniquely due to it being developed for a large device. Then, we broke it down into two views and a fullscreen mode. Afterward, we implemented the first view, which was the highlight view. To do this, we figured out the components required. We then implemented all the components using SwiftUI. At the end of the previous chapter, we only had a fancy wireframe for the highlight view and no connection to another view. Now, we will create the enhanced view and implement all the functionality to provide navigation between the views and send image metadata between the highlight and enhanced views.

This chapter will be split into the following sections:

- EnhancedView Design Overview
- Updating HighlightView
- Testing EnhancedView
- Extra tasks

By the end of this chapter, you will have created a fully functional photo gallery that leverages the iPad's immense screen real estate, which is yours to be modified, tweaked, and used as you see fit. I will give you exercises when we reach the end of the chapter to implement more advanced functionality into the photo gallery. This will transition nicely into our next project, the Mac App Store.

Technical Requirements

This chapter requires you to download Xcode version 14 or above from Apple's App Store.

To install Xcode just search for Xcode in the App Store, then select and download the latest version. Open Xcode and follow any additional installation instructions. Once Xcode has opened and launched, you're ready to go.

Version 14 of Xcode has the following features/requirements:

- Includes SDKs for iOS 16, iPadOS 16, macOS 12.3, tvOS 16, and watchOS 9.

- Supports on-device debugging in iOS 11 or later, tvOS 11 or later, and watchOS 4 or later.

- Requires a Mac running macOS Monterey 12.5 or later.

For further information regarding technical details, please refer to *Chapter 1*.

The code files for this chapter can be found here:

```
https://github.com/PacktPublishing/Elevate-SwiftUI-Skills-by-Building-
Projects/tree/main/Code/Chapter%205%20-%20iPad%20Project%20-%20
Photo%20Gallery%20Enhanced%20View
```

In the next section, we will look at the EnhancedView. We will break it down into components that we can implement.

EnhancedView design overview

In this section, we will implement the EnhancedView. If you recall, in the previous chapter, we discussed the design of the EnhancedView. The following figures show the portrait and landscape modes of the EnhancedView as a reminder:

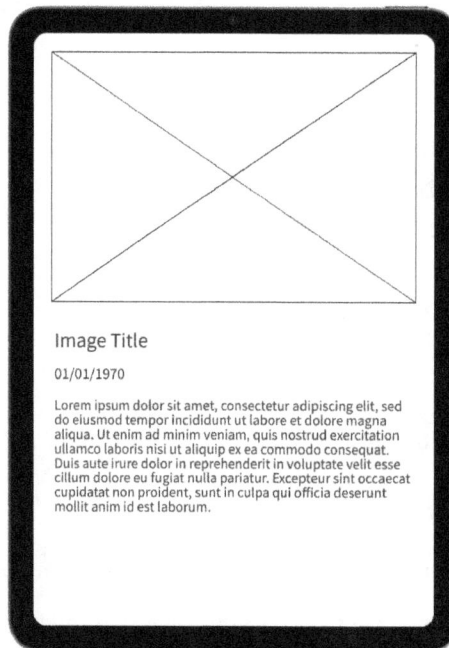

Figure 5.1 – EnhancedView wireframe preview in portrait mode

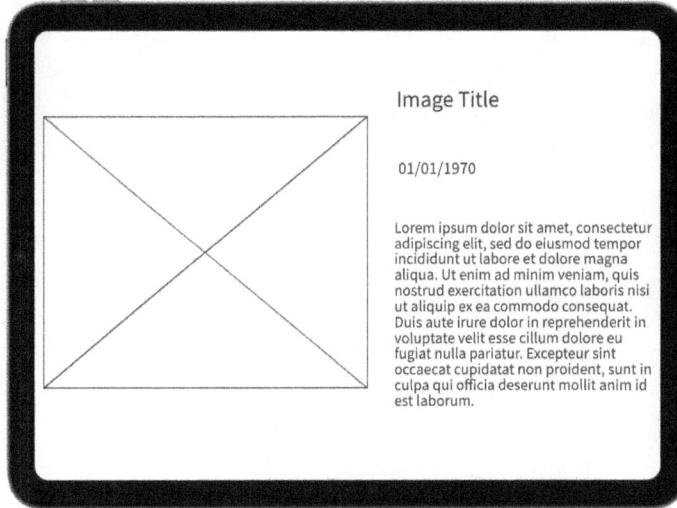

Figure 5.2 – EnhancedView wireframe preview in landscape mode

Before we code our application, we will break down the `EnhancedView` into the elements that comprise it. As a little task, see whether you can figure out what they are. Don't worry if you don't know the exact UI component names; we will look at the components in the following sections.

> **Important note**
> The components are the same for portrait and landscape orientation.

The Text component

The `Text` component is one of the simplest components offered by SwiftUI. It allows you to display a string of characters/numbers, which is very useful for headings and providing information. We will use it three times for the following features:

- Image title
- Date
- Image description

The following figures show Text components:

Image Title

Figure 5.3 – Image Title text component

01/01/1970

Figure 5.4 – Image date text component

Lorem ipsum dolor sit amet, consectetur adipiscing elit, sed do eiusmod tempor incididunt ut labore et dolore magna aliqua. Ut enim ad minim veniam, quis nostrud exercitation ullamco laboris nisi ut aliquip ex ea commodo consequat. Duis aute irure dolor in reprehenderit in voluptate velit esse cillum dolore eu fugiat nulla pariatur. Excepteur sint occaecat cupidatat non proident, sunt in culpa qui officia deserunt mollit anim id est laborum.

Figure 5.5 – Image description text component

The Image component

The Image component is one of the core components offered by SwiftUI. It allows you to display an image, which can be used to provide a visual representation or to a body of text. We will use it to show a bigger version of the image that was selected from the HighlightView. The following figure shows the image in EnhancedView:

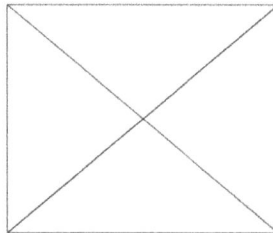

Figure 5.6 – Image component

In the next section, we will create EnhancedView and implement it using the components we discussed using SwiftUI in our application.

Adding EnhancedView Components

In this section, we will add the previously discussed components to create our EnhancedView. However, we first need to create the EnhancedView file. Doing so is simple; follow these steps:

1. We will now create a new SwiftUI View for the results page. Right-click the gallery folder inside of your **Project Navigator** pane and select **New File…**:

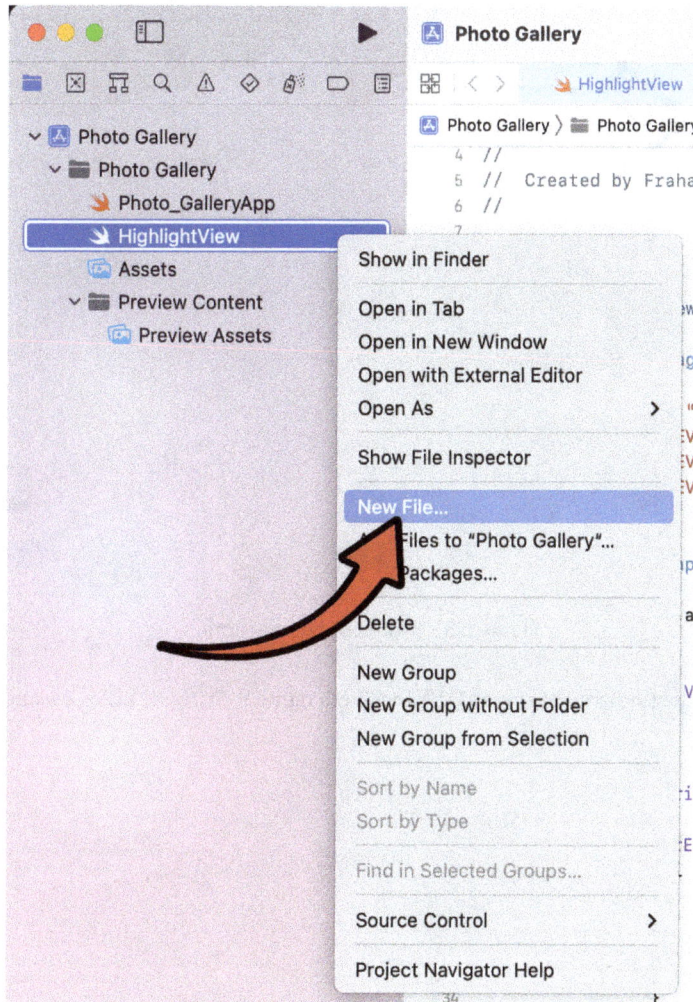

Figure 5.7 – Select New File…

2. Next, we will select the type of file we want to add, which is a **SwiftUI View** (selecting this provides a SwiftUI template, which saves us the time and effort of retyping the SwiftUI file structure every time) in the **User Interface** section:

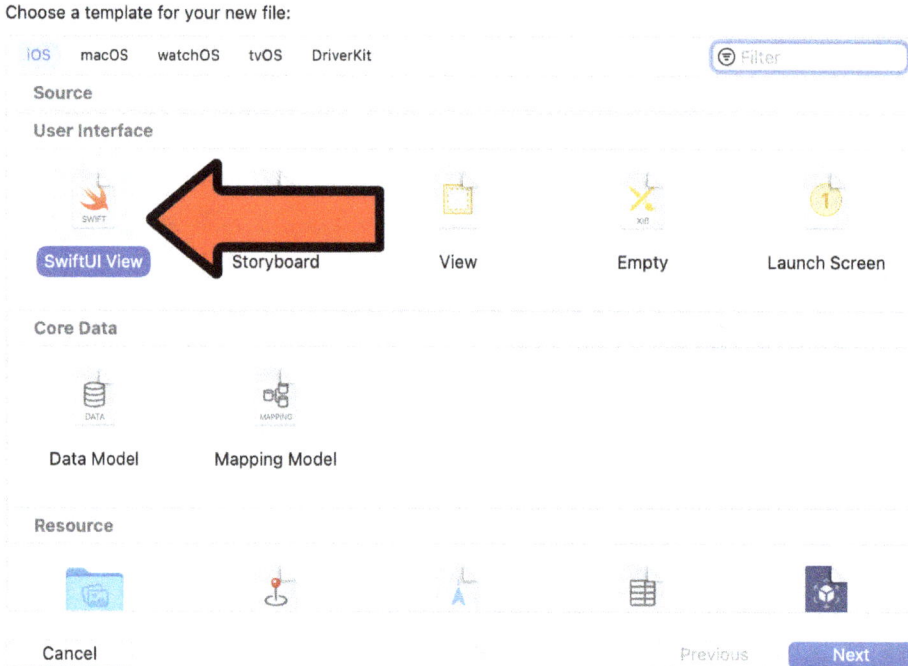

Choose a template for your new file:

iOS   macOS   watchOS   tvOS   DriverKit   🔽 Filter

Source

User Interface

SwiftUI View   Storyboard   View   Empty   Launch Screen

Core Data

Data Model   Mapping Model

Resource

Cancel   Previous   Next

Figure 5.8 – SwiftUI View selection

3. Finally, we must rename our **SwiftUI View**. Let's name it `EnhancedView` and press **Create**:

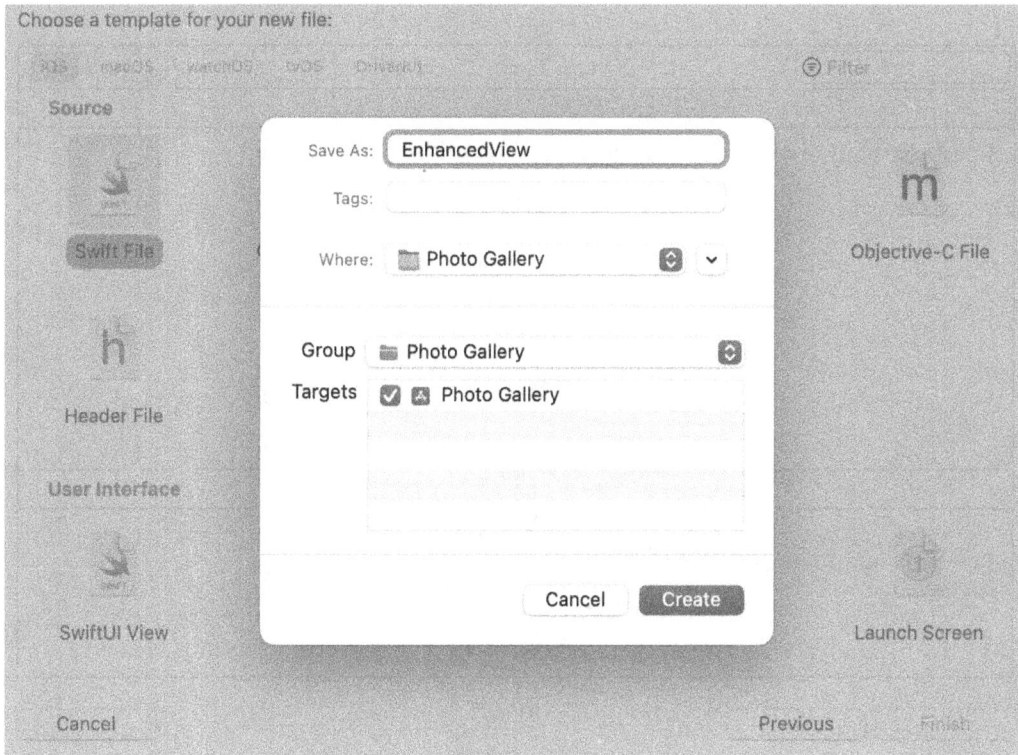

Figure 5.9 – Naming our view

So far, we have looked at the wireframes and the components that compose them. Finally, we created our EnhancedView file.

Updating HighlightView

The first thing I will do is update the code to align with my coding standards; feel free to do the same. Now, we will add five state variables. One will be used to track whether an image has been clicked, and the other four will be used to pass data from the HighlightView to the EnhancedView. Add the code after the adaptive columns like so:

```
private let adaptiveColumns =
[
    GridItem( .adaptive( minimum: 300 ) )
]

@State private var isClicked: Bool = false
@State private var imageFile: String = ""
```

```
@State private var imageName: String = ""
@State private var imageDate: String = ""
@State private var imageDescription: String = ""
```

Now, update the body to match the following code:

```
var body: some View
{
    NavigationView
    {
        ScrollView
        {
            LazyVGrid( columns: adaptiveColumns, spacing: 20 )
            {
                ForEach( images.indices )
                { i in
                    NavigationLink( destination: EnhancedView(
imageFile: $imageFile, imageName: $imageName, imageDate: $imageDate,
imageDescription: $imageDescription ), isActive: $isClicked, label:
                    {
                        Image( images[i] )
                            .resizable( )
                            .scaledToFill( )
                            .frame( width: 300, height: 300 )
                            .onTapGesture {
                                imageFile = images[i]
                                imageName = "FireDEV Podcast"
                                imageDate = "22/09/2022"
                                imageDescription = "Aspiring
entrepreneurs and industry professionals alike can learn a lot from
a fireside chat with interesting people in the industry. From small
indie developers to CEOs of major companies, these chats provide
an opportunity to gain insight into the unique stories of success
that have led these individuals to their current positions. Through
conversations about their experiences and challenges, we can gain
valuable knowledge about their successes, failures, and the strategies
they used to reach their goals. We can also gain insight into their
motivations and the values that drive their decisions. By engaging
in a fireside chat with these industry leaders, we can gain a better
understanding of the industry and the people within it, and gain
valuable knowledge that can help us to reach our own goals."

                                isClicked = true
                    }
                } )
            }
```

```
            }
        }
    }
    .navigationViewStyle( StackNavigationViewStyle( ) )
}
```

We just added a lot of code, so let's unpack it all:

- `NavigationView {`: We implement a navigation system that will allow us to go to the `EnhancedView` and back. Using this code for the `NavigationView` alone will result in a split system, thus at the end we add the proceeding code to remove the split view mode.

- `.navigationViewStyle( StackNavigationViewStyle( ) )`: Coupled with the new `NavigationView`, we wrap each image around a `NavigationLink`, which allows it to be clickable so we can navigate to another view.

- `NavigationLink( destination: EnhancedView( imageFile: $imageFile, imageName: $imageName, imageDate: $imageDate, imageDescription: $imageDescription ), isActive: $isClicked, label:` Within the `NavigationLink`, we have several parameters. Let's break down the purpose of each one:

  - `destination: EnhancedView(`: Sets the view to navigate to upon clicking the image.

  - `imageFile: $imageFile`: Passes the `imageFile` state variable that was created previously to the `EnhancedView`. This variable is the path/filename of the image. This will be stored as a string.

  - `imageName: $imageName`: Passes the `imageName` state variable to the `EnhancedView`. This variable is the name/title of the image.

  - `imageDate: $imageDate`: Passes the `imageDate` state variable that was created previously to the `EnhancedView`. This variable is the date of the image. This could be the creation date, edit date, or some other relevant date.

  - `imageDescription: $imageDescription`: Passes the `imageDescription` state variable to the `EnhancedView`. This variable is the description of the image.

- `isActive: $isClicked`: Tracks whether the image has been clicked. If so, it will navigate to the `EnhancedView`.

- `label`: Although we are not using any form of text for the button, the image will be used as a label, which will act as the clickable label/image for navigation.

The next step is to add clickable functionality to the image allowing the user to navigate from the `HighlightView` to the `EnhancedView`. Update the image code within the `NavigationLink` as follows:

```
Image( images[i] )
    .resizable( )
    .scaledToFill( )
    .frame( width: 300, height: 300 )
    .onTapGesture {
        imageFile = images[i]
        imageName = "FireDEV Podcast"
        imageDate = "22/09/2022"
        imageDescription = "Aspiring entrepreneurs and industry
professionals alike can learn a lot from a fireside chat with
interesting people in the industry. From small indie developers
to CEOs of major companies, these chats provide an opportunity to
gain insight into the unique stories of success that have led these
individuals to their current positions. Through conversations about
their experiences and challenges, we can gain valuable knowledge about
their successes, failures, and the strategies they used to reach their
goals. We can also gain insight into their motivations and the values
that drive their decisions. By engaging in a fireside chat with these
industry leaders, we can gain a better understanding of the industry
and the people within it, and gain valuable knowledge that can help us
to reach our own goals."

        isClicked = true
    }
```

We have added an `onTapGesture` function. The purpose of this is to assign the metadata for the image to the `@State` variables created earlier. All the metadata variables are hardcoded, except `imageFile`, which uses an array. Feel free to extend the current array to become a multidimensional data container to store unique metadata for each image. Finally, we set `isClicked` to `true`; this tells the view to navigate to the designated view on the `NavigationLink`, which is the `EnhancedView`.

All these changes will result in the following code for the `HighlightView`:

```
import SwiftUI

struct HighlightView: View
{
    private let images: [String] =
    [
        "FireDEV", "FireDEV", "FireDEV", "FireDEV", "FireDEV",
"FireDEV", "FireDEV", "FireDEV", "FireDEV", "FireDEV", "FireDEV",
"FireDEV", "FireDEV", "FireDEV", "FireDEV", "FireDEV", "FireDEV",
"FireDEV", "FireDEV", "FireDEV"
    ]
```

```
    private let adaptiveColumns =
    [
        GridItem( .adaptive( minimum: 300 ) )
    ]

    @State private var isClicked: Bool = false
    @State private var imageFile: String = ""
    @State private var imageName: String = ""
    @State private var imageDate: String = ""
    @State private var imageDescription: String = ""


    var body: some View
    {
        NavigationView
        {
            ScrollView
            {
                LazyVGrid( columns: adaptiveColumns, spacing: 20 )
                {
                    ForEach( images.indices )
                    { i in
                        NavigationLink( destination: EnhancedView(
imageFile: $imageFile, imageName: $imageName, imageDate: $imageDate,
imageDescription: $imageDescription ), isActive: $isClicked, label:
                        {
                        Image( images[i] )
                            .resizable( )
                            .scaledToFill( )
                            .frame( width: 300, height: 300 )
                            .onTapGesture {
                                imageFile = images[i]
                                imageName = "FireDEV Podcast"
                                imageDate = "22/09/2022"
                                imageDescription = "Aspiring
entrepreneurs and industry professionals alike can learn a lot from
a fireside chat with interesting people in the industry. From small
indie developers to CEOs of major companies, these chats provide
an opportunity to gain insight into the unique stories of success
that have led these individuals to their current positions. Through
conversations about their experiences and challenges, we can gain
valuable knowledge about their successes, failures, and the strategies
they used to reach their goals. We can also gain insight into their
motivations and the values that drive their decisions. By engaging
in a fireside chat with these industry leaders, we can gain a better
understanding of the industry and the people within it, and gain
valuable knowledge that can help us to reach our own goals."
```

```
                                    isClicked = true
                                }
                        } )
                    }
                }
            }
        }
        .navigationViewStyle( StackNavigationViewStyle( ) )

    }
}

struct ContentView_Previews: PreviewProvider
{
    static var previews: some View
    {
        HighlightView( )
    }
}
```

That was a lot to unpack! Feel free to take another look at this section before moving on. Remember, you have access to the GitHub repository for online access for easy copy and paste: `https://github.com/PacktPublishing/Elevate-SwiftUI-Skills-by-Building-Projects`.

Implementing EnhancedView

First, we need to implement code to handle the rotation functionality. Add the following code above the `EnhancedView` struct:

```
struct DeviceRotationViewModifier: ViewModifier
{
    let action: ( UIDeviceOrientation ) -> Void

    func body( content: Content ) -> some View
    {
        content
            .onAppear( )
            .onReceive( NotificationCenter.default.publisher( for:
UIDevice.orientationDidChangeNotification ) )
                { _ in
                    action( UIDevice.current.orientation )
                }
    }
```

```
}

extension View
{
    func onRotate( perform action: @escaping ( UIDeviceOrientation )
-> Void ) -> some View
        {
            self.modifier( DeviceRotationViewModifier( action: action
) )
        }
}
```

The preceding code we added allows us to render the content again when rotating the screen. This will be used shortly when detecting the device's orientation. Now, we can add the @Binding variables at the start of our EnhancedView struct, which allows us to pass in metadata:

```
@Binding var imageFile: String
@Binding var imageName: String
@Binding var imageDate: String
@Binding var imageDescription: String
```

Now, we will add two variables, the first detecting the device's orientation and the second detecting the device screen size in pixels. The former will be used to determine the correct layout, and the latter will be used when sizing components. Add the following code:

```
@State private var orientation = UIDeviceOrientation.unknown
let screenSize: CGRect = UIScreen.main.bounds
```

In the body, we will create a group, which will contain two layouts, one for each orientation: portrait and landscape. Add the following code, and we will discuss everything that is happening:

```
Group
{
    if ( orientation.isLandscape )
    {
        LazyHStack
        {
            VStack
            {
                Image( imageFile )
                    .resizable( )
                    .scaledToFit( )
            }.frame( width: screenSize.width * 0.5 )

            VStack
```

```
                    {
                        Text( imageName )
                            .fontWeight(.bold)
                        Text( imageDate )

                        Text( imageDescription )
                    }.frame( width: screenSize.width * 0.5 )
                }
            }
        else
        {
            LazyVStack
            {
                VStack
                {
                    Image( imageFile )
                        .resizable( )
                        .scaledToFit( )
                }.frame( height: screenSize.height * 0.5 )

                VStack
                {
                    Text( imageName )
                        .fontWeight( .bold )
                    Text( imageDate )

                    Text( imageDescription )
                }.frame( height: screenSize.height * 0.5 )
                }
            }
        }
    .onRotate
    { newOrientation in
        orientation = newOrientation
    }
```

Let's run through the code we just added, bit by bit. First, we check which orientation the device is in. By default, we check whether it's landscape. If not, it must be portrait, and we handle the component sizes and positioning accordingly:

```
if ( orientation.isLandscape )
```

Next, we create a lazy horizontal stack for storing the components in the landscape orientation:

```
LazyHStack
```

Next, we create two vertical stacks. The frame width is set to half of the screen's width. This effectively creates an equal splitscreen design. Feel free to modify the multiplier if you would like a custom split. Inside the first vertical stack, we put the image, which is set to `resizable` and `scaledToFit`. In the second vertical stack, we put the text metadata in simple `Text` components:

```
VStack
{
    Image ( imageFile )
        .resizable ( )
        .scaledToFit ( )
}.frame ( width: screenSize.width * 0.5 )

VStack
{
    Text ( imageName )
        .fontWeight (.bold)
    Text ( imageDate )

    Text ( imageDescription )
}.frame ( width: screenSize.width * 0.5 )
```

In the `else` statement, we simply use `LazyVStack` as it pertains to the portrait orientation. The only other change made is to the frame size: it is not linked to the screen's width but its height. The rest remains the same.

Finally, we add a detector to our `Group` component, which simply detects when the device has been rotated and updates the orientation variable, which is used to detect which mode to draw:

```
.onRotate
{ newOrientation in
    orientation = newOrientation
}
```

The only thing that remains is updating the preview provider at the bottom of the `EnhancedView` file. Update the code as follows:

```
struct EnhancedView_Previews: PreviewProvider
{
    static var previews: some View
    {
        EnhancedView ( imageFile: .constant ( "" ), imageName:
```

```
.constant( "" ), imageDate: .constant( "" ), imageDescription:
.constant( "" ) )
    }
}
```

The preceding code just adds a default set of parameters to be passed through for the preview. I have left it blank as I used the simulator for testing, but feel free to put in dummy data to ensure you can properly add text using the preview.

In this section, we implemented the EnhancedView code. In the next section, we will take a look at the result. Feel free to modify the layout for landscape and portrait to make it unique.

Testing EnhancedView

In this section, we will finally get to test our application. Launching it up will take us to the HighlightView; click on any image and it will take you to the EnhancedView. The portrait mode will look as follows:

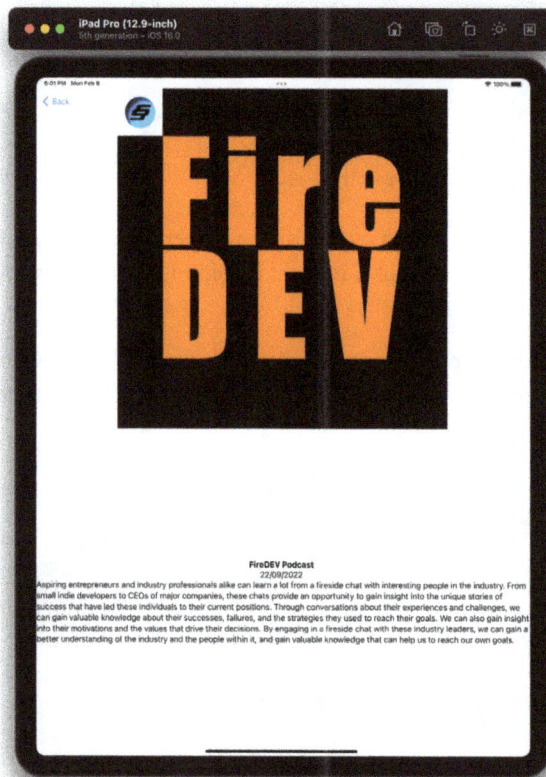

Figure 5.10 – Portrait mode

Rotating the application will result in the following output:

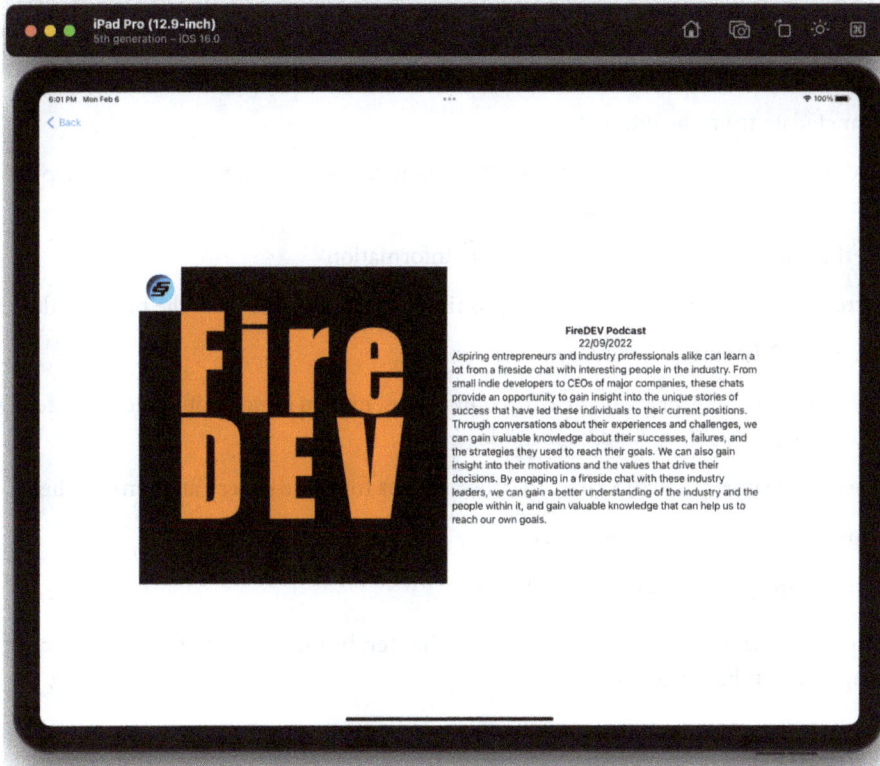

Figure 5.11 – Landscape mode

Now, our application is complete and features a navigational menu for going back to the `HighlightView`.

> **Note**
> If you require help with rotating the simulator, please refer to the previous chapter.

Extra tasks

Now that the application is complete, here is a list of extra tasks for you to complete to enhance your application:

- Use different source data:

  - Different images
  - Different title

- Different description

- Different date

- Load images from the internet

- Load metadata from the internet

- Extend the scope of supported devices to iPhone as well, thus providing you with the opportunity to consider cross-platform design

- Make the image fullscreen without any extra information

- Fullscreen tap for more info: A single tap while in fullscreen will show the photo's title

- Collections: Different sets of images

- A side panel, which displays all the collection names, hidden in portrait mode, activated using a button; always visible in landscape mode

- Delete and rename: Allows the user to delete images from the gallery and rename them

- Sharing: The ability to share an image

- Different display modes: List and grid view

We will summarize what we have covered in this chapter, but first, we will look at the code for implementing a few of the extra tasks.

Fullscreen mode

In order to add fullscreen mode to the `EnhancedView`, we will add a new `@State` variable called `isFullScreen`. We will use this variable to toggle between fullscreen mode and regular mode. Additionally, we will need to add `onTapGesture` to the image so that when the image is tapped, it toggles fullscreen mode. Here is the modified code:

```
import SwiftUI

struct EnhancedView: View
{
    @Binding var imageFile: String
    @Binding var imageName: String
    @Binding var imageDate: String
    @Binding var imageDescription: String

    @State private var orientation = UIDeviceOrientation.unknown
    @State private var isFullScreen: Bool = false

    let screenSize: CGRect = UIScreen.main.bounds
```

```
var body: some View
{
    Group
    {
        if isFullScreen {
            Image(imageFile)
                .resizable()
                .scaledToFit()
                .edgesIgnoringSafeArea(.all)
                .onTapGesture {
                    self.isFullScreen.toggle()
                }
        }
        else if orientation.isLandscape {
            LazyHStack
            {
                VStack
                {
                    Image(imageFile)
                        .resizable()
                        .scaledToFit()
                        .onTapGesture {
                            self.isFullScreen.toggle()
                        }
                }.frame(width: screenSize.width * 0.5)

                VStack
                {
                    Text(imageName)
                        .fontWeight(.bold)
                    Text(imageDate)

                    Text(imageDescription)
                }.frame(width: screenSize.width * 0.5)
            }
        }
        else {
            LazyVStack
            {
                VStack
                {
                    Image(imageFile)
```

```
                            .resizable()
                            .scaledToFit()
                            .onTapGesture {
                                self.isFullScreen.toggle()
                            }
                    }.frame(height: screenSize.height * 0.5)

                    VStack
                    {
                        Text(imageName)
                            .fontWeight(.bold)
                        Text(imageDate)

                        Text(imageDescription)
                    }.frame(height: screenSize.height * 0.5)
                }
            }
        }
    }
}

struct EnhancedView_Previews: PreviewProvider
{
    static var previews: some View
    {
        EnhancedView(imageFile: .constant(""), imageName:
.constant(""), imageDate: .constant(""), imageDescription:
.constant(""))
    }
}
```

Let's see what this modified code does:

- Adds a new @State variable, isFullScreen, to keep track of whether the view is in fullscreen mode or not.

- Adds a new condition at the beginning of the Group that, if isFullScreen is true, shows the image in fullscreen mode. In this condition, we use .edgesIgnoringSafeArea(. all) to ensure the image takes up the entire screen, and .onTapGesture to toggle isFullScreen when the image is tapped.

- Modifies the existing Image views in both landscape and portrait mode by adding .onTapGesture to toggle isFullScreen when the image is tapped.

This results in the image taking up the full screen when tapped and returning to its original size when tapped again.

Collections

To add collections (albums) and a side panel to display the names of these collections, we will make several changes to the code:

1. Create a data structure to represent a collection of images.

2. Modify `HighlightView` to display a list of collections in a side panel when in landscape mode.

3. Show the images of the selected collection.

This is how you can do it:

```
import SwiftUI

// Data structure representing an image collection
struct ImageCollection {
    let name: String
    let images: [String]
}

struct HighlightView: View {
    // Sample data
    private let collections: [ImageCollection] = [
        ImageCollection(name: "Collection 1", images: ["FireDEV",
"FireDEV", "FireDEV"]),
        ImageCollection(name: "Collection 2", images: ["FireDEV",
"FireDEV", "FireDEV", "FireDEV"]),
        ImageCollection(name: "Collection 3", images: ["FireDEV",
"FireDEV"])
    ]

    private let adaptiveColumns = [GridItem(.adaptive(minimum: 300))]

    @State private var isClicked: Bool = false
    @State private var imageFile: String = ""
    @State private var imageName: String = ""
    @State private var imageDate: String = ""
    @State private var imageDescription: String = ""
    @State private var selectedCollection: ImageCollection?

    var body: some View {
        NavigationView {
```

```swift
GeometryReader { geometry in
    if geometry.size.width > geometry.size.height {
        // Horizontal mode, show side panel
        HStack {
            // Side Panel
            List(collections, id: \.name) { collection in
                Button(action: {
                    selectedCollection = collection
                }) {
                    Text(collection.name)
                }
            }
            .frame(width: geometry.size.width * 0.25)

            // Images
            ScrollView {
                LazyVGrid(columns: adaptiveColumns,
spacing: 20) {
                    if let images = selectedCollection?.
images {
                        ForEach(0..<images.count, id:
\.self) { i in
                            Image(images[i])
                                .resizable()
                                .scaledToFill()
                                .frame(width: 300, height:
300)
                        }
                    }
                }
            }
        }
    } else {
        // Vertical mode, just show images
        ScrollView {
            LazyVGrid(columns: adaptiveColumns, spacing:
20) {
                if let images = selectedCollection?.images
{
                    ForEach(0..<images.count, id: \.self)
{ i in
                        Image(images[i])
                            .resizable()
                            .scaledToFill()
                            .frame(width: 300, height:
```

```
300)
                                }
                            }
                        }
                    }
                    .onAppear {
                        // Select the first collection by default
                        if selectedCollection == nil {
                            selectedCollection = collections.first
                        }
                    }
                }
            }
        }
        .navigationViewStyle(StackNavigationViewStyle())
    }
}

struct ContentView_Previews: PreviewProvider {
    static var previews: some View {
        HighlightView()
    }
}
```

Let's explain the changes:

- We added a struct called ImageCollection that represents a collection of images with a name.
- We updated the collections property to be an array of ImageCollections.
- We removed the old images array as it's now part of the collections.
- We used GeometryReader to determine whether the view is in horizontal or vertical mode. In horizontal mode, a side panel is displayed with a list of collection names.
- In horizontal mode, clicking on a collection name in the side panel updates the selectedCollection state variable, which in turn updates the images displayed to the right of the side panel.
- In vertical mode, only the images of the selected collection are displayed. By default, the first collection is selected.

This code demonstrates how you can create an adaptive layout that shows a side panel in horizontal mode and adjusts its content based on the selected collection.

Summary

In this chapter, we covered the design of our `EnhancedView` using wireframes. These wireframes helped us break down the views into their components. We then implemented the SwiftUI components to match the design from the wireframe. Though the components were the same for the portrait and landscape orientation, we configured their positions and sizes accordingly. It is very important to make sure each orientation that is supported is best utilized in line with industry standards. We also updated the `HighlightView` to pass in data to the `EnhancedView`. This data was used to display content in components added in `EnhancedView`. Then, we covered extra tasks for you to undertake; feel free to look over this chapter again before proceeding. We have now completed our second application, which is ready for you to modify and use as you see fit.

In our next chapter, we will start our next application, the App Store for Mac. We will naturally look at the design and break it down to help us understand and implement the application for our next platform.

6

Mac Project – App Store Bars

In the previous four chapters, we created a tax calculator app for the iPhone and a photo gallery app for the iPad. We implemented them from scratch, looking at the technical requirements, design specifications, wireframes, and code implementation. We will use the skills covered in this and the next chapter to create the App Store sidebar, but worry not: we will go over all necessary aspects in case you have jumped straight to this chapter.

In this chapter, we will work on the design of our third project, an App Store application for the Mac that will showcase its big screen, which also stays in one orientation. We will assess the requirements for designing such an application and discuss the design specifications, allowing us to get a better understanding of what is required and how it will all fit together. Then, we will start our application's coding process to build the sidebars in this chapter, and the next chapter will cover the main body of the App Store app. This project will cover the foundations of SwiftUI components.

This chapter covers the following main topics:

- Understanding the Design Specifications
- Acceptance Criteria
- Building the Sidebar UIs

By the end of this chapter, you will have a better understanding of the design of our application and what is required, along with a skeleton user interface that will be used as the foundation for making the gallery work in the next chapter.

Technical requirements

This chapter requires you to download Xcode version 14 or above from Apple's App Store. To install Xcode, just search for Xcode in the App Store, then select and download the latest version. Open Xcode and follow any additional installation instructions. Once Xcode has opened and launched, you're ready to go.

Version 14 of Xcode has the following features/requirements:

- Includes SDKs for iOS 16, iPadOS 16, macOS 12.3, tvOS 16, and watchOS 9

- Supports on-device debugging in iOS 11 or later, tvOS 11 or later, and watchOS 4 or later

- Requires a Mac running macOS Monterey 12.5 or later

For further information regarding technical details, please refer to *Chapter 1*.

The code files for this chapter can be found here: `https://github.com/PacktPublishing/Elevate-SwiftUI-Skills-by-Building-Projects`

In the next section, we will provide clarity on the specifications of our application's design and look at mockups of what the app will look like.

Understanding the Design Specifications

In this section, we will look at the design specifications of our App Store application and describe the features we are going to implement in it. The best method for figuring out the features required is to put yourself in the user's shoes to determine how they will use the app, then break it into individual steps.

There can be many features of our app, and they are as follows:

- Sidebar – for different sections of the app.

- Highlight banner – showcasing an app of the day.

- Featured tiles – showcasing lesser apps or more specific category applications.

- New apps and updates section.

- Search – the ability to search the whole collection for a specific app

- Account management.

- App reviewing.

- App reporting.

- App page – showing information such as description and date for a selected app, similar to the `EnhancedView` from the photo gallery application we created previously.

- App images and information from an external source such as a local database or online.

- Sharing – the ability to share an image.

- Creating favorite/download later lists.

Now that we have listed the ideal features we would like, next, it is important for us to determine which features are absolutely crucial. To do this, we must understand the end use of our product. For me, the purpose of creating this App Store application is to showcase a sidebar and how it integrates with the main body. We will not be implementing an app page because it is very similar to the photo gallery's `EnhancedView`, which will be set as an extra task, and you can use the previous two chapters for assistance. Based on that, I know that not all the features are required, and some would be useful if omitted and assigned as extra tasks for you as the developer to undertake. Therefore, the following are the core features we will be implementing:

- Sidebar – for different sections of the app.
- Highlight banner – showcasing an app of the day.
- Featured tiles – showcasing lesser apps or more specific category applications.
- New apps and updates section.

The rest of the features will be an exercise for you once you have completed this and the next chapter. The next section will cover the acceptance criteria for our application.

Understanding the Acceptance criteria

We will discuss the mandatory requirements for our application that we absolutely want to see in the end product at the end of the next chapter. If possible, we should try to make them measurable, so let's list them right now:

- Sidebar with text components and an icon next to them for enhanced visual context.
- Scrollable main body view to scroll through the applications and highlighted applications in the main body.
- Highlight banner that will take the full width of its parent container to showcase an application.
- Tiles to showcase other specific apps in a category.
- Images and text displayed for the rest of the apps.

Develop test cases in which the application's acceptance criteria will be tested. Using this method allows the ability to see the conditions in which the application will be used by the end user and the level that needs to be attained for it to be considered successful.

We will develop testing methods to check the acceptance criteria that can be tested and ultimately measured. This will allow us to see whether the use cases in which the application is to be used will pass.

Wireframe design

One of the most useful tools for designing layouts is wireframing. A wireframe is an overview of how the layout will look. The following figure shows the whole front page of the application that we will be implementing using a wireframe:

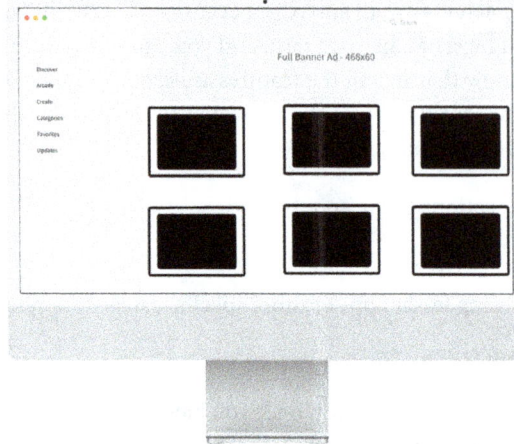

Figure 6.1 – App Store wireframe

In the next section, we will build the interface for our application and make sure it looks the way we designed it in the wireframe. Though we will build it the same way, there can be small differences. We will focus on the sidebar in this chapter and complete the main body in the next chapter.

Building the Sidebar UI

We will now build the UI for the sidebar. First, let's create our project. Follow the following steps:

1. Open Xcode and select **Create a new Xcode project**:

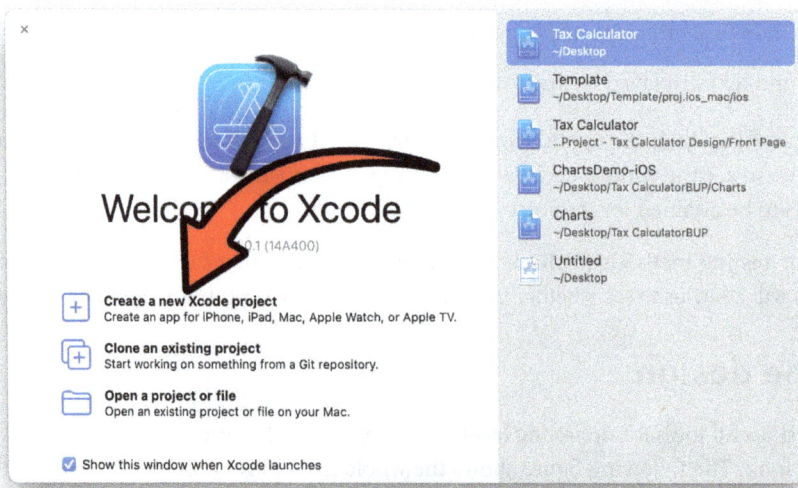

Figure 6.2 – Create a new Xcode project

2. Now, we will choose the template for our application. As we are creating an iPad app, we will select **iOS** from the top, select **App**, and click **Next**:

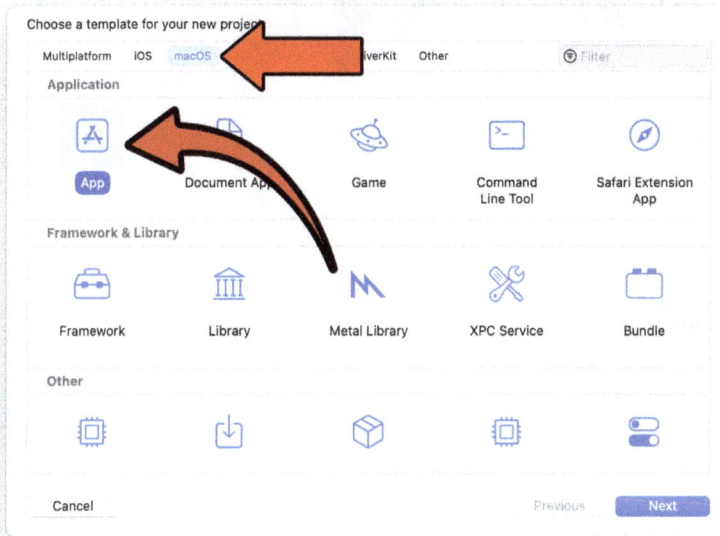

Figure 6.3 – Xcode project template selection

3. We will now choose the options for our project. Here, there are only two crucial things to select/set. Make sure that the interface is set to **SwiftUI**, which will be the UI our system will use, and that **Language** is set to **Swift**, which is obviously the programming language used for our application:

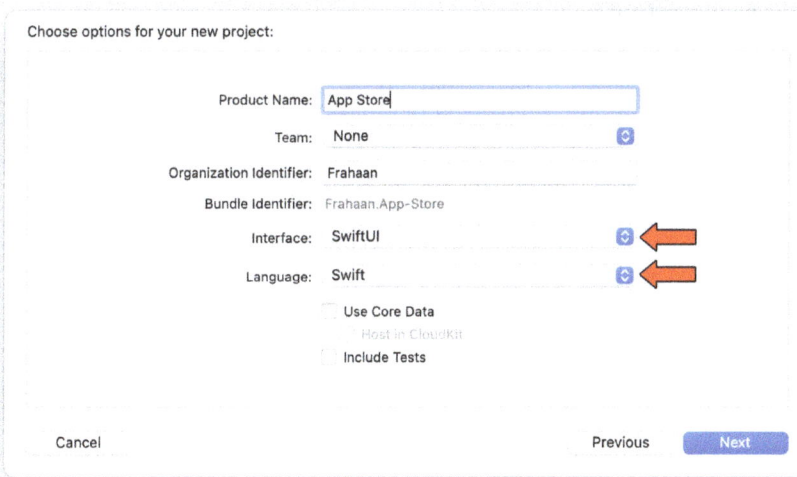

Figure 6.4 – Xcode project options

4.   Once you press **Next**, you can choose where to create your project, as seen in the following figure:

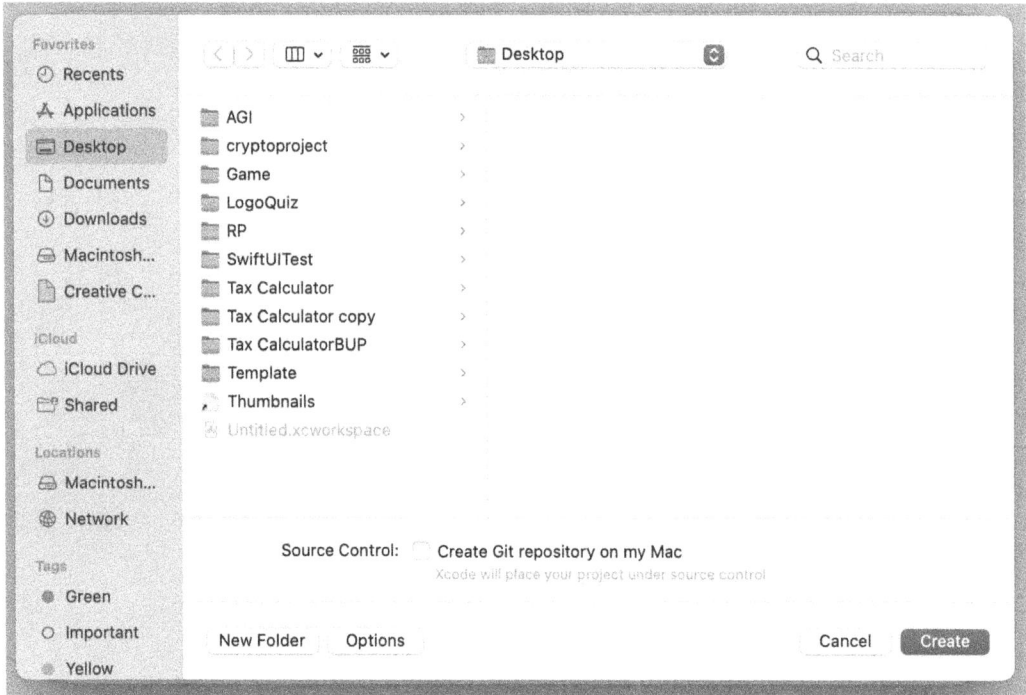

Figure 6.5 – Xcode project save directory

5.   Once you have found the location you would like to create it in, go ahead and click on **Create** in the bottom-right corner as seen in *Figure 6.5*. Xcode shows your project in all its glory, as seen in the following figure:

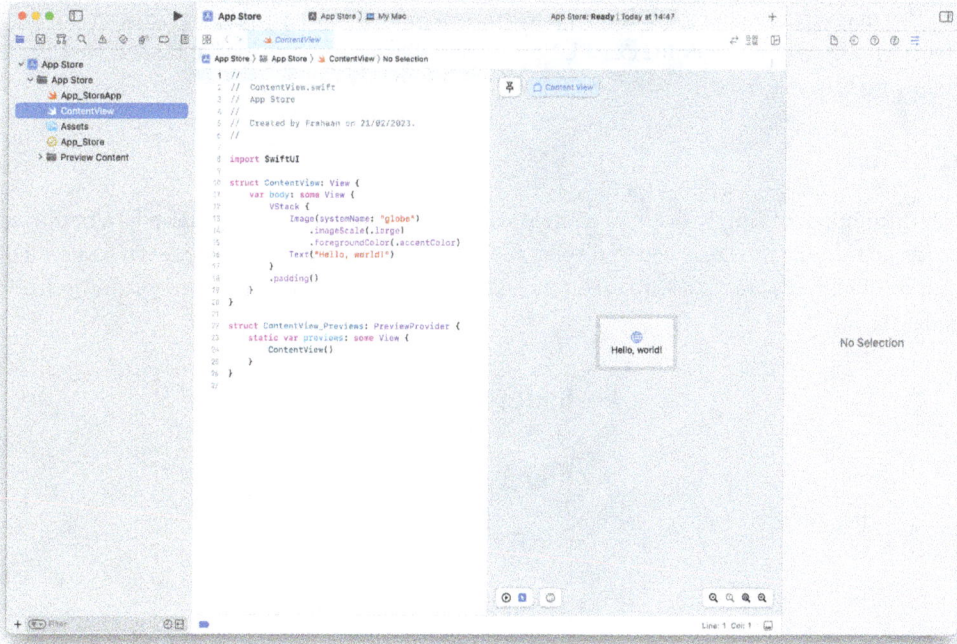

Figure 6.6 – New Xcode project overview

In the next section, we will implement the sidebar of our application using SwiftUI.

Exploring the Sidebar components

In this section, we will implement the sidebar's user interface. As a reminder, it will look like the following:

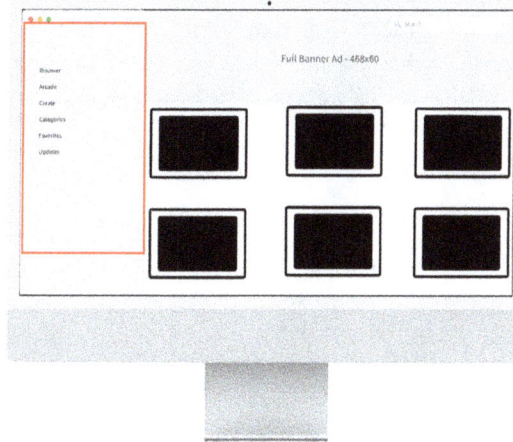

Figure 6.7 – App Store wireframe

There are two main elements to the sidebar. As a little task, see whether you can figure out what they are. Don't worry if you don't know the exact UI components' names; we will look at these components in the following sections.

Label item

A label item component simply displays an item within the sidebar that can be used as a button to navigate the application. It allows you to display a string of characters, numbers, or even icons, all of which can be used in conjunction with each other. For us, we will use them as dummy buttons inside of our sidebar:

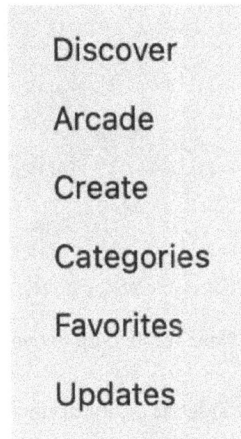

Discover

Arcade

Create

Categories

Favorites

Updates

Figure 6.8 – Sidebar item

SearchBar

A SearchBar component allows the user to search through a set list of components. For us, we will use it as a dummy search component that will search through all the apps in the App Store. Though the search bar isn't part of the sidebar per se, we will implement it along with the sidebar:

Q Search

Figure 6.9 – Search bar

Renaming the view

In this section, we will add the image components to our highlight page, which is currently named **ContentView**. Firstly, let's rename `ContentView` to `MainView`. If you already know how to rename it, feel free to skip these steps. Doing this is simple: open `ContentView`, right-click any reference to `ContentView` in the code, then go to **Refactor | Rename...**, as shown in the following screenshot:

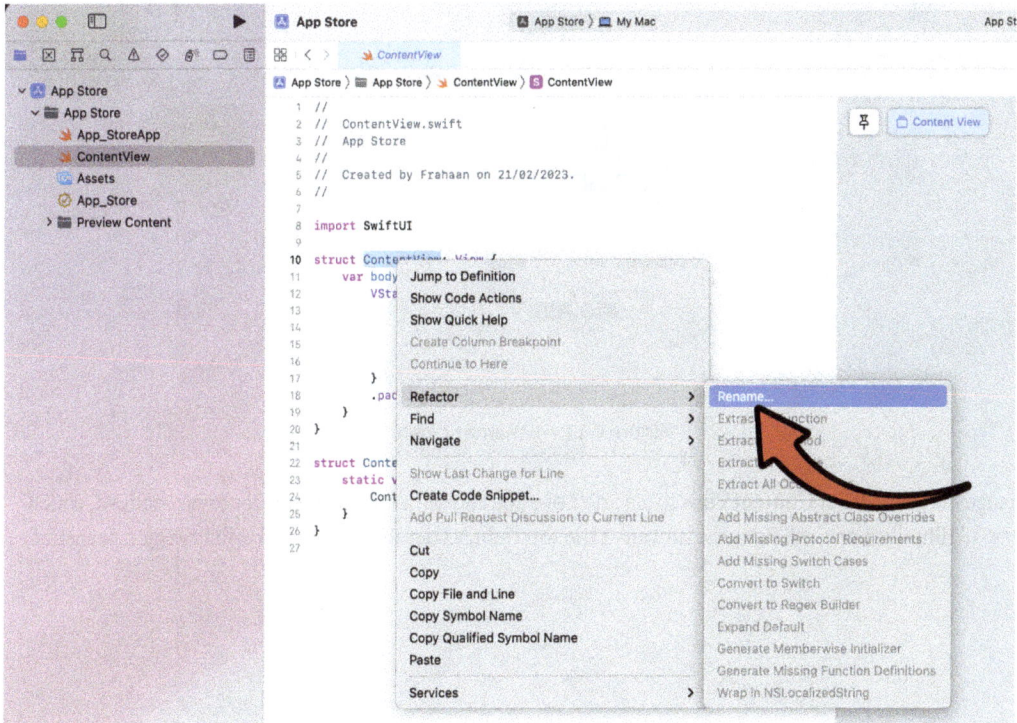

Figure 6.10 – Rename button

Next, the following screen will be shown:

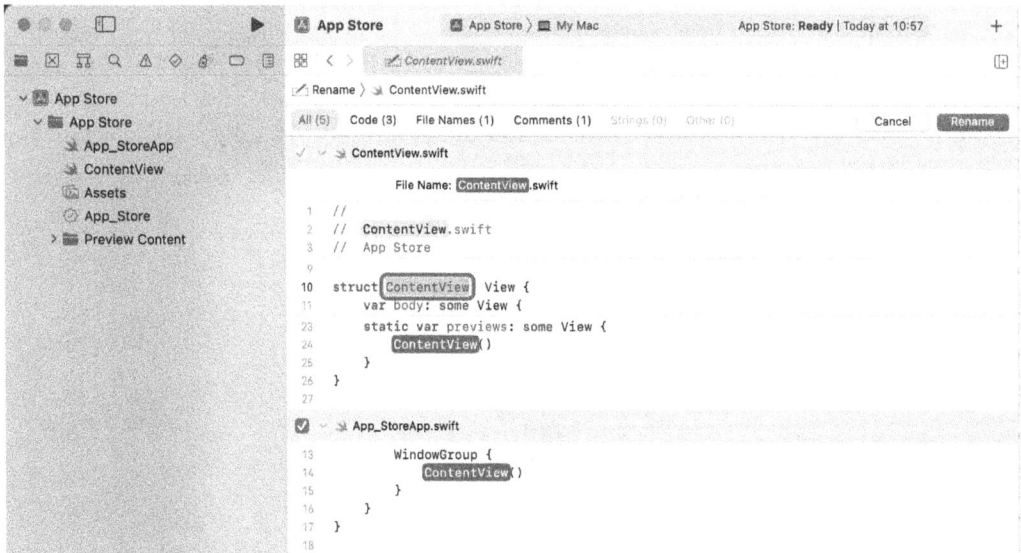

Figure 6.11 – Rename screen

Change the name from **ContentView** to MainView. You can see all other references to be changed, which is useful. Finally, press the **Rename** button at the top right, as demonstrated in the following screenshot:

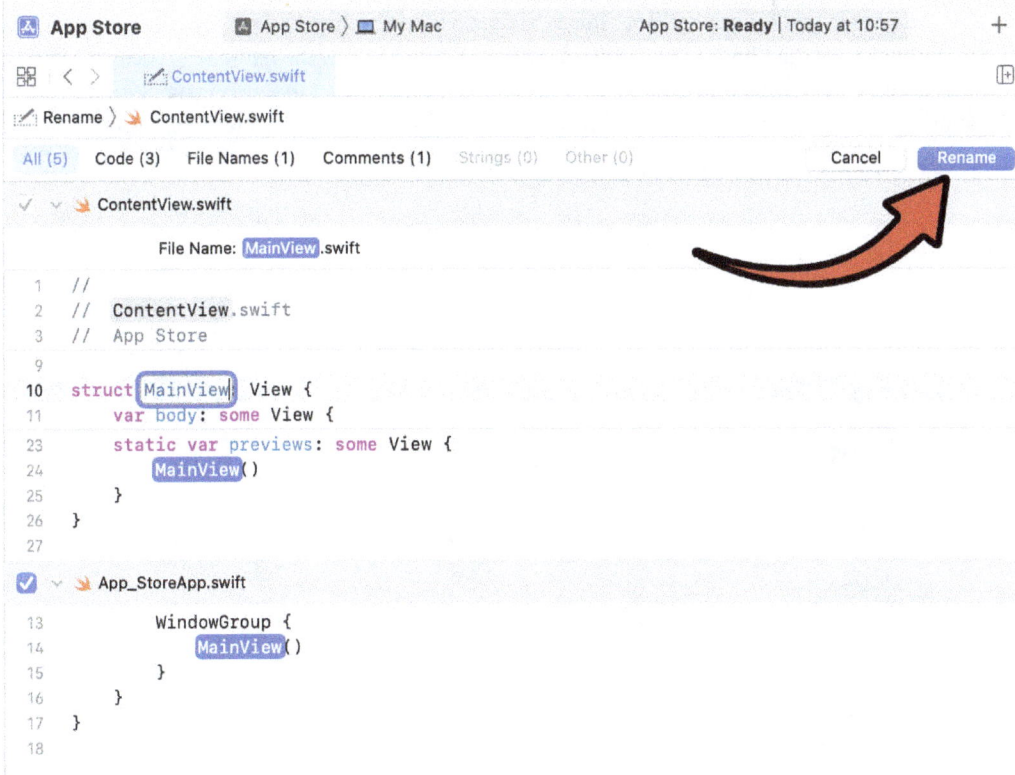

Figure 6.12 – Rename button

We have now renamed our view, including the file, as can be seen in **Project Navigator**:

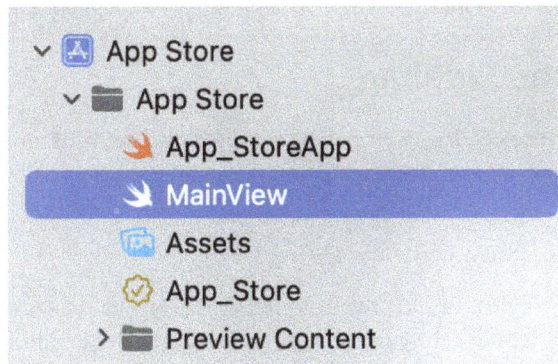

Figure 6.13 – Updated filename in Project Navigator

There is one extra step that is optional. That step is to rename the `ContentView_Previews` struct. Though not crucial, I would highly recommend renaming it to keep all the name references in sync. Using the preceding steps, rename the `ContentView_Previews` struct to `MainView_Previews`. The location of the struct is at the bottom of the `MainView` file (previously named `ContentView`):

```swift
 8  import SwiftUI
 9
10  struct MainView: View {
11      var body: some View {
12          VStack {
13              Image(systemName: "globe")
14                  .imageScale(.large)
15                  .foregroundColor(.accentColor)
16              Text("Hello, world!")
17          }
18          .padding()
19      }
20  }
21
22  struct ContentView_Previews: PreviewProvider {
23      static var previews: some View {
24          MainView()
25      }
26  }
27
```

Figure 6.14 – Rename Previews struct

In this section, we took a look at the design for the application and, more specifically, the `SideBar` UI. We also renamed `ContentView`. In the next section, we will implement the code for the sidebar.

Implementing the Sidebar

As we have created a fresh project, the coding standards aren't in line with my personal preference. So, firstly, I will change the standards. Feel free to take a few moments to do the same.

NavigationView

Let's implement a `NavigationView` to provide a split-screen layout, which will allow us to code a sidebar. Doing so is simple. Remove the current body code and replace it with the following:

```swift
var body: some View
{
    NavigationView
```

```
{
    List
    {
        Label( "Discover", systemImage: "" )
        Label( "Arcade", systemImage: "" )
        Label( "Create", systemImage: "" )
        Label( "Categories", systemImage: "" )
        Label( "Favorites", systemImage: "" )
        Label( "Updates", systemImage: "" )
    }
}
}
```

Within our `NavigationView`, we create a `List` that has a collection of `Label`s that will serve as our dummy buttons. By default, we provide no system image, but we need to specify something otherwise it will result in an error.

The preceding code will result in the following:

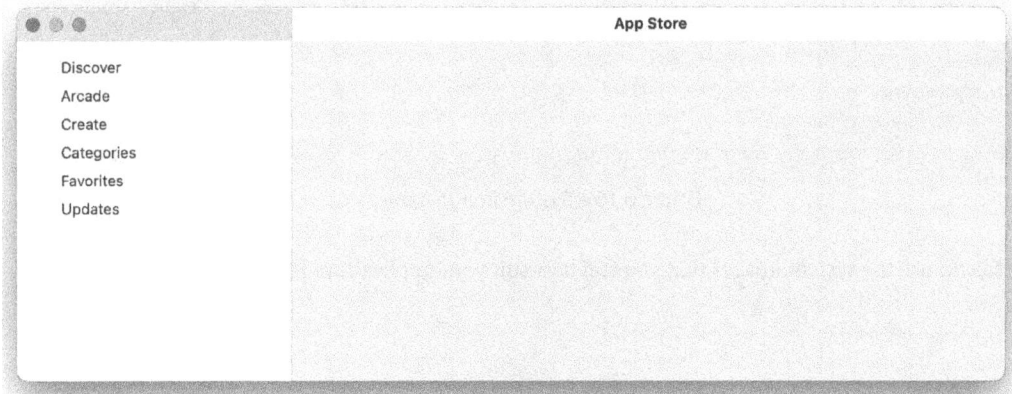

Figure 6.15 – Labels preview

Though we do not need to implement icons, the application will greatly benefit from their inclusion. Adding an image is easy enough; you simply use the `systemImage` parameter. Alternatively, you can provide your own images by following the steps from *Chapter 4* in the *Implementing the Highlight View* section and implementing them using the `image` parameter. The best method for finding/searching for system images is to download **SF Symbols** from Apple, which you can find at `https://developer.apple.com/sf-symbols/`. Once that's installed, you can search for system images easily. To use a system image, simply copy the following name of the system image:

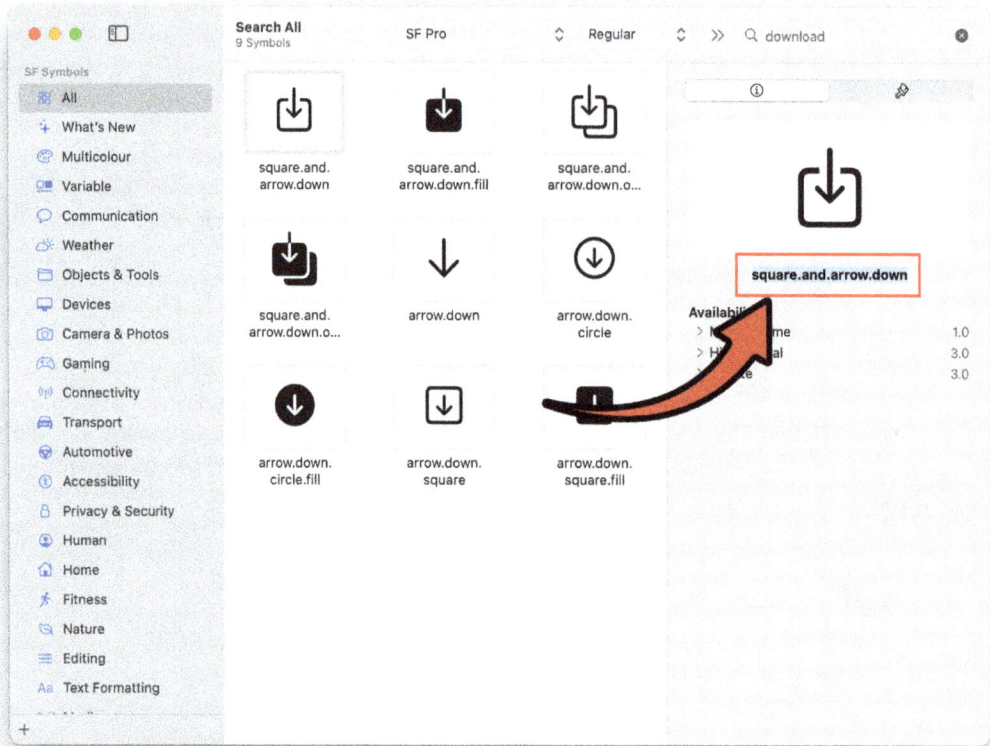

Figure 6.16 – System image name

Feel free to use the system images that you feel best suit your application. I have chosen the following:

```
List
{
    Label( "Discover", systemImage: "star" )
    Label( "Arcade", systemImage: "gamecontroller" )
    Label( "Create", systemImage: "paintbrush" )
    Label( "Categories", systemImage: "square.grid.3x3.square" )
    Label( "Favorites", systemImage: "heart" )
    Label( "Updates", systemImage: "square.and.arrow.down" )
}
```

Running the application will display the following result:

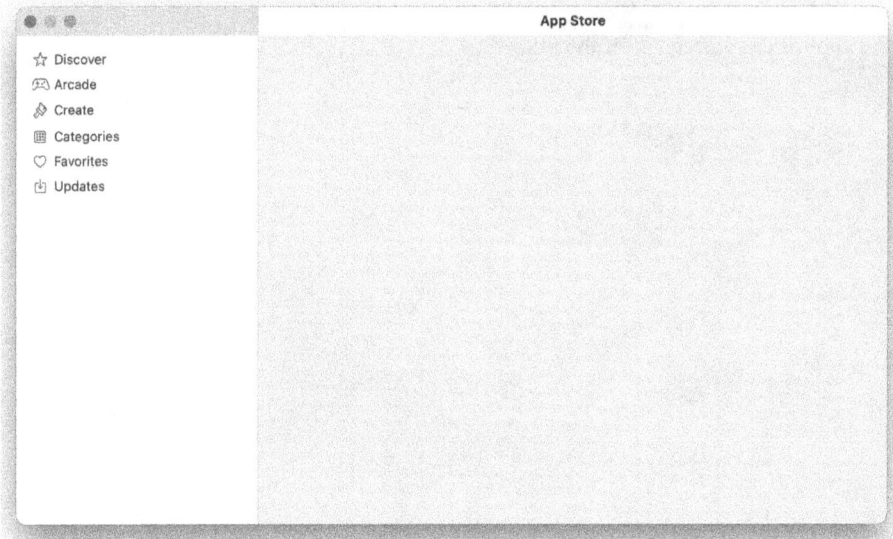

Figure 6.17 – System image icons

SearchBar

Though the `SearchBar` will be placed in the top bar, it is included in the sidebar code. Adding the search bar is very simple. First, add the following code before the body, which will store the user input of the `SearchBar`:

```
@State private var searchText = ""
```

Next, add the following code to the `List` component:

```
.searchable( text: $searchText )
```

The code we have just added adds a `SearchBar` and links its text to the `searchText` variable we created. The list component code will now look as follows:

```
List
{
    Label( "Discover", systemImage: "star" )
    Label( "Arcade", systemImage: "gamecontroller" )
    Label( "Create", systemImage: "paintbrush" )
    Label( "Categories", systemImage:
"square.grid.3x3.square" )
    Label( "Favorites", systemImage: "heart" )
```

```
    Label( "Updates", systemImage: "square.and.arrow.down" )
}.searchable( text: $searchText )
```

Running the application will result in the following:

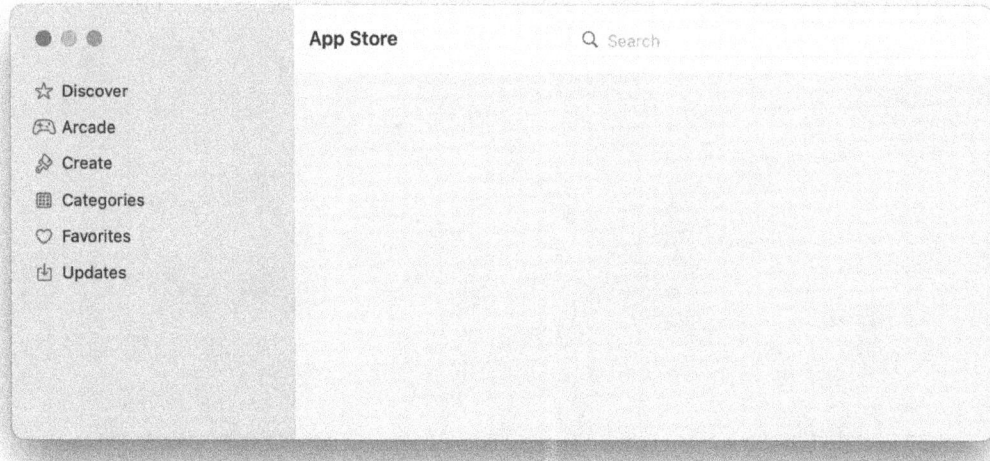

Figure 6.18 – SearchBar

In this section, we implemented the code for the `SearchBar` UI. In the next section, we will look at extra features that will help bring the App Store to life.

Implementing Extra Features

Even though for the scope of this project we are done with the `SideBar`, I would like to show you how to implement events for pressing *Enter* on the `SearchBar` and how to make the `SideBar` labels clickable.

SearchBar Enter Event

We want the user to be able to press *Enter* when they have selected `SearchBar` and trigger an event. This event could pull up a list of results in a context menu or new page. Feel free to implement this as an extra task.

The code to achieve this is super simple. Add the following after the `.searchable` code:

```
}.searchable( text: $searchText )
    .onSubmit( of: .search )
    {
        print( searchText )
    }
```

When the user has submitted the search, the code inside of the parentheses will run. For testing, we are printing out the searchText variable, which will print out what you type in it. Feel free to run it.

Clickable Label Event

Right now, the labels inside the sidebar have no event functionality. For the scope of this chapter, we don't require it, but I will show you how to make the labels clickable. It is actually very simple. Just add an onTapGesture function to each label as follows:

```
Label( "Discover", systemImage: "star" )
.onTapGesture
{
    print( "Discover" )
}
```

I have only added it to the first label, but feel free to add it to the rest.

In this section, we implemented code to allow the user to trigger the Submit event when pressing *Enter* in the search bar. We also added the code to detect when the labels have been clicked.

Summary

In this chapter, we covered the design of our App Store application. We looked at wireframes and broke down each element into SwiftUI components. We then implemented the SwiftUI components to match the design from the wireframes for the SideBar UI. We also took a look at the requirements for building this application and the design specifications. We then simplified it to the core features that our app will provide. We also implemented extra features to allow extra input for a more well-rounded application.

In the next chapter, we'll take a look at implementing the main body for our App Store application.

7

Mac Project – App Store Body

In this chapter, we will work on implementing the main body for the App Store project. In the previous chapter, we looked at the design of the App Store and more specifically the SideBar design. Then, we broke the SideBar down into all the necessary components required for our application requirements. We then implemented all the components using SwiftUI. At the end of the previous chapter, we only had a SideBar with some optional event tracking, but no content in the main body. The main section will be scrollable and showcase apps using icons and banners. Now we will analyze the main body, break it down into all the components it comprises, and implement all the components to provide an app store-like feel.

This chapter will be split into the following sections:

- Main body overview
- Implementing the main body
- Extra Tasks

By the end of this chapter, you will have created an app store template with a scrollable view to showcase applications. This will serve as a solid foundation for further expanding the app store application or pivoting the project to something totally different while using the core structure we have implemented. As we reach the end of the chapter, I will give exercises to implement more advanced functionality in the app store. This will transition nicely into our fourth and final project, the Apple Watch Fitness Companion App.

Technical Requirements

This chapter requires you to download Xcode version 14 or above from Apple's App Store.

To install Xcode, just search for Xcode in the App Store, select, and download the latest version. Open Xcode and follow any additional installation instructions. Once Xcode has opened and launched, you're ready to go.

Version 14 of Xcode has the following features/requirements:

- Includes SDKs for iOS 16, iPadOS 16, macOS 12.3, tvOS 16, and watchOS 9

- Supports on-device debugging in iOS 11 or later, tvOS 11 or later, and watchOS 4 or later

- Requires a Mac running macOS Monterey 12.5 or later

For further information regarding the technical details, please refer to *Chapter 1*.

The code files for this chapter can be found here:

```
https://github.com/PacktPublishing/Elevate-SwiftUI-Skills-by-Building-
Projects
```

In the next section, we will provide clarity on the specifications of our application's design and look at mockups of what the app will look like.

Main body overview

In this section, we will take another look at the wireframes from *Chapter 6* and break them down into their individual components. The wireframe images for the app store and the main body have been provided in this section. These images depict the layout and design of the app store and the main body.

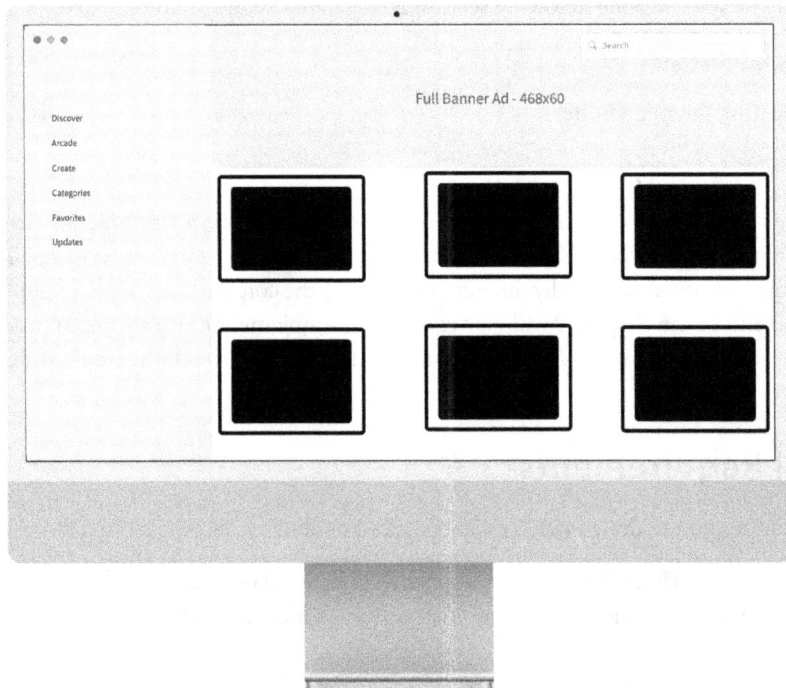

Figure 7.1 – App Store view

Full Banner Ad - 468x60

Figure 7.2 – App Store main body

Before we code our application, we will break down the main body into the elements that comprise it. As a little task, see whether you can figure out what these are, but don't worry if you don't know the exact UI component names. We will look at the components in the following section.

Image components

An Image component is one of the core components offered by SwiftUI. It allows you to display an image, which can be used to provide a visual representation or to aid a body of text. We will use it in two main ways, firstly to showcase a particular app using a Highlight banner, and secondly to show a list/grid of applications. The following figures show the application icon and application highlight banner:

Figure 7.3 – App icon

Full Banner Ad - 468x60

Figure 7.4 – Banner

Text component

The Text component is one of the simplest components offered by SwiftUI. It allows you to display a string of characters/numbers, which is very useful for headings and providing information. We will use it for the following:

- App title

- Section description

In the upcoming section, we will proceed to develop the main body of our application using the SwiftUI components that we previously discussed. This implementation will be carried out with utmost precision and attention to detail.

Implementing the main body

In this section, we will complete the third project in this book by implementing the main body of our application. Our first step will be to code the Highlight banner, followed by the app icons.

Coding the highlight banner

Firstly, we will add the code for a Highlight banner. The banner is simply going to be an image that spans the width of the body; we will give it some spacing for aesthetic purposes. It is common to add multiple banners throughout the page to highlight different applications and have carousel banners, which provide the ability to showcase multiple banners in a single section through a transition such as sliding. We will implement a single banner; however, adding more is simple. Follow these steps:

1. Let's start by adding a banner image. My image is **728x90** pixels. Feel free to modify this to suit your needs. Select **Assets** from the **Project Navigator**:

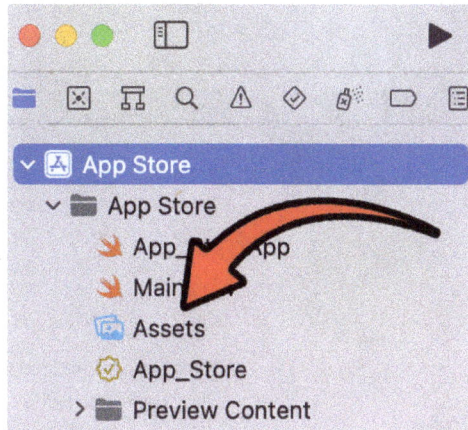

Figure 7.5 – Assets location in Project Navigator

2. Now, the **Assets** view will appear. Importing an image into **Assets** can be done in one of two ways

    I.   Dragging and dropping the files into the **Assets** section:

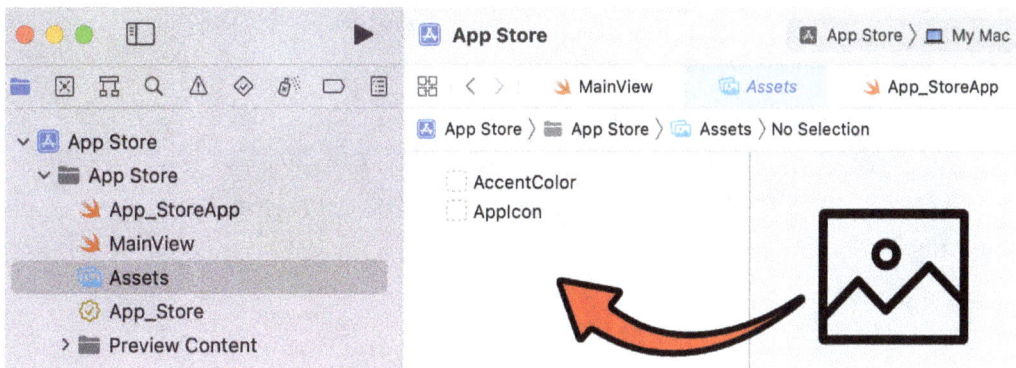

Figure 7.6 – Dragging and dropping asset

    II.   Right-clicking the **Assets** section and selecting **Import…**:

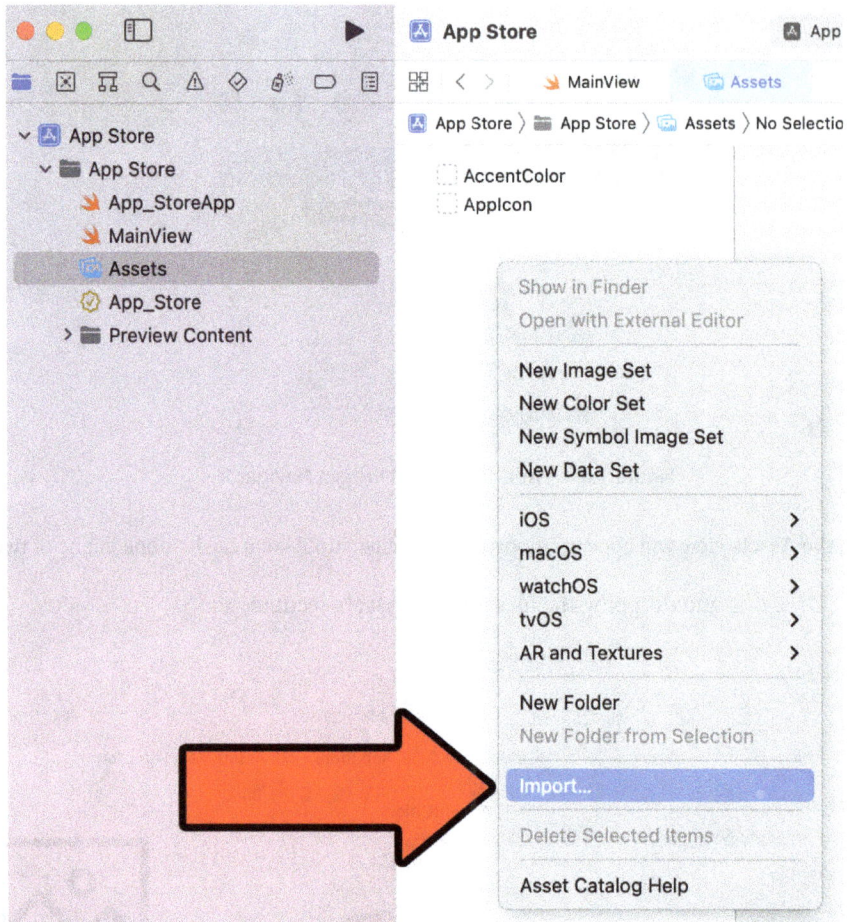

Figure 7.7 – Import… button

Once the asset(s) have been imported, the **Assets** view will look as follows:

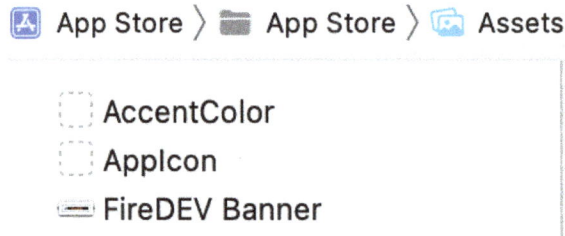

Figure 7.8 – Asset(s) imported

> **Note**
>
> I am using the banner for my developer-centric podcast FireDEV. Feel free to use it and tune in to my podcast every Thursday at the following links:
>
> - **Spotify**: `https://open.spotify.com/show/387RiHksQE33KYHTitFXhg`
> - **Apple Podcasts**: `https://podcasts.apple.com/us/podcast/firedev-fireside-chat-with-industry-professionals/id1602599831`
> - **Google Podcasts**: `https://podcasts.google.com/feed/aHR0cHM6Ly9hbmNob3IuZm0vcy83Yjg2YTNiNC9wb2RjYXN0L3Jzcw`

3. We will add an `Image` component after the `List` code:

```
}.searchable( text: $searchText )
    .onSubmit( of: .search )
    {
        print( searchText )
    }
Image( "FireDEV Banner" )
```

This will result in the following:

Figure 7.9 – Banner added

4. If you try and resize the window, there are restrictions. We need to make the banner resizable and maintain its original aspect ratio. Update the image code as follows:

```
Image( "FireDEV Banner" )
    .resizable( )
    .padding( .horizontal )
    .scaledToFit( )
```

We made the image resizable, which allows it to change size based on the size of the window. This is very useful as the user will run the App Store on different screen sizes and may not always have it fullscreen. We then added horizontal padding to make sure it doesn't touch the left or right edges. This can be omitted if you like, or you can specify a set amount of padding. Finally, we set it to `scaledToFit`, which maintains the original aspect ratio. Distortion is never a good idea. All of this results in the following:

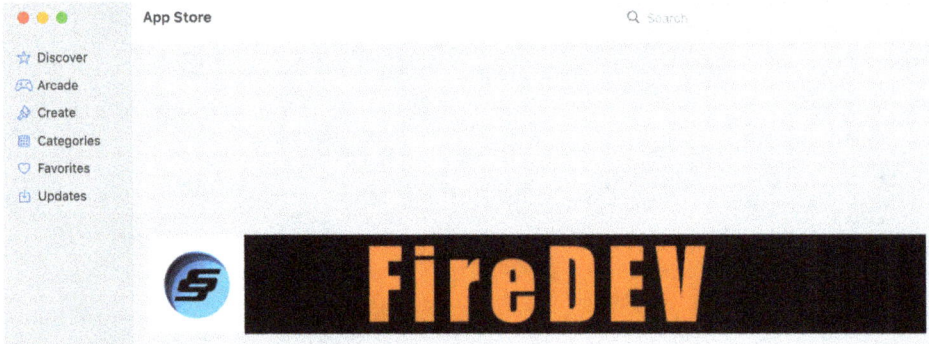

Figure 7.10 – Banner updated

5. As of now, the banner is always in the center. We want it at the top of the view. To achieve this result, we will enclose the image code we added previously within a `ScrollView` with an alignment of `topLeading`. Update the code like so:

```
ScrollView
{
    Color.clear

    Image( "FireDEV Banner" )
        .resizable( )
        .padding( .horizontal )
        .scaledToFit( )
}
```

We also add a `Color.clear` instruction to make sure there is no background color, all of which results in the following awesome banner:

Figure 7.11 – Banner positioned at the top

The highlight banner has been finished, which can be converted into a carousel. Moving forward, the app groups will be coded to showcase a list of application icons.

Coding the app groups

We will now implement the code to display app groups. These will contain an image representing the application icon and a label that represents the application name. Feel free to add more components to each group and arrange them as you see fit. I have added an app icon to the assets. I followed the previous steps to add images. Feel free to refer to those steps:

1. First, add the following code before the body:

```
private let adaptiveColumns =
[
    GridItem( .adaptive( minimum: 300 ) )
]
```

This will be used in our grid and ensures the items have a minimum size of 300. This is extremely useful as we don't want them to become so small the user cannot see them.

2. Add the following code beneath the banner code we added in the previous section:

```
LazyVGrid( columns: adaptiveColumns, spacing: 20 )
{
    ForEach ( 0..<20 )
    { index in
        VStack( alignment: .leading )
        {
            Image( "FireDEV Logo" )
            Label( "FireDEV", systemImage: "" )
        }
    }
}
```

Let's break down the code before we run the application.

- We create a `LazyVGrid` using `adaptiveColumns` and with `spacing` set to `20`. Feel free to change the spacing of the column sizing as you see fit.

- Next, we use a `ForEach` loop that runs 20 times. Feel free to substitute the code with an array of items as done in the previous project.

- Then, we create a **vertical stack**, which will contain the app group. We use a vertical stack as it allows us to put the `Label` component beneath the image. The label will be used as the name of the application.

- Finally, we create an `Image` component and a `Label` component, and we omit the `systemImage` parameter as it is not required by us. However, you must put something, hence the empty quotation marks.

Running the application will result in the following:

Figure 7.12 – App groups

3. We are almost done with this application. The application name is a little on the small side. Let's make it bigger, and update `Label` as follows:

```
Label( "FireDEV", systemImage: "" )
    .font( .system( size: 36 ) )
```

Finally, running this will result in the following figure:

Figure 7.13 – Label font size increase

As a recap, here is the whole code for `MainView`:

```
//
//  ContentView.swift
//  App Store
//
//  Created by Frahaan on 21/02/2023.
//

import SwiftUI

struct MainView: View
{
    @State private var searchText = ""

    private let adaptiveColumns =
    [
        GridItem( .adaptive( minimum: 300 ) )
    ]

    var body: some View
    {
```

```
        NavigationView
        {
            List
            {
                Label( "Discover", systemImage: "star" )
                .onTapGesture
                {
                    print( "Discover" )
                }
                Label( "Arcade", systemImage: "gamecontroller" )
                Label( "Create", systemImage: "paintbrush" )
                Label( "Categories", systemImage: "square.
grid.3x3.square" )
                Label( "Favorites", systemImage: "heart" )
                Label( "Updates", systemImage: "square.and.
arrow.down" )
            }.searchable( text: $searchText )
            .onSubmit( of: .search )
            {
                print( searchText )
            }


            ScrollView
            {
            Color.clear

            Image( "FireDEV Banner" )
                .resizable( )
                .padding( .horizontal )
                .scaledToFit( )

            LazyVGrid( columns: adaptiveColumns, spacing: 20
)
            {
                ForEach ( 0..<20 )
                { index in
                    VStack( alignment: .leading )
                    {
                        Image( "FireDEV Logo" )
                        Label( "FireDEV", systemImage: "" )
                            .font( .system( size: 36 ) )
                    }
                }
```

```
                    }
                }
            }
        }
    }

struct MainView_Previews: PreviewProvider
{
    static var previews: some View
    {
        MainView( )
    }
}
```

The code is also available in the GitHub repository of the book.

In this section, we implemented the main body for our App Store application, thus finishing our third project. There were two main sections – first, we implemented a Highlight Banner, which can be used multiple times throughout the view to showcase different applications. Then, we implemented a grid of app groups. Although the app information was hardcoded, it can be abstracted into an array, which can store more information for each application. In the next section, we will summarize this chapter.

Extra Tasks

Now that the application is complete, here is a list of extra tasks for you to complete to enhance your application:

- An app page, to which the user navigates when they click an app icon
- Multiple highlight banners
- Updating the banner to be a carousel
- Different pages for the section in the sidebar
- Pulling data from an array or external source such as a database for the app metadata

In the next section, we will summarize what we have covered in this chapter, but first, we will look over the code to help with the extra tasks for this project.

Search Functionality

To add search functionality to the app, you can use the `.searchable` modifier provided by SwiftUI. Here's the modified code with search functionality added:

```
//
//  ContentView.swift
```

```
// App Store
//
// Created by Frahaan on 21/02/2023.
//

import SwiftUI

struct MainView: View {
    @State private var searchText = ""

    private let adaptiveColumns = [
        GridItem(.adaptive(minimum: 300))
    ]

    var body: some View {
        NavigationView {
            List {
                Label("Discover", systemImage: "star")
                    .onTapGesture {
                        print("Discover")
                    }
                Label("Arcade", systemImage: "gamecontroller")
                Label("Create", systemImage: "paintbrush")
                Label("Categories", systemImage: "square.grid.3x3.
square")
                Label("Favorites", systemImage: "heart")
                Label("Updates", systemImage: "square.and.arrow.down")
            }
            .searchable(text: $searchText) // Add searchable modifier
            .onSubmit(of: .search) {
                print(searchText)
            }

            ScrollView {
                Color.clear

                Image("FireDEV Banner")
                    .resizable()
                    .padding(.horizontal)
                    .scaledToFit()

                LazyVGrid(columns: adaptiveColumns, spacing: 20) {
                    ForEach(0..<20) { index in
                        VStack(alignment: .leading) {
```

```
                                Image("FireDEV Logo")
                                Label("FireDEV", systemImage: "")
                                    .font(.system(size: 36))
                            }
                        }
                    }
                }
            }
        }
    }
}

struct MainView_Previews: PreviewProvider {
    static var previews: some View {
        MainView()
    }
}
```

In this modified code, I added the `.searchable(text: $searchText)` modifier to the `List` view, which enables search functionality. The `searchText` variable is used as the binding for the search text input.

I also added an `.onSubmit(of: .search)` modifier to the `List` view to handle the search submission. In this example, it prints the search text to the console, but you can customize the action based on your requirements.

With these modifications, users will be able to enter search queries and filter the items in the list based on the entered text.

App Page

To provide an enhanced page with more information when the user clicks on an app, you can create a new view that displays detailed information about the selected app. Here's an example of how you can modify the code to achieve this:

```
import SwiftUI

struct AppDetailsView: View {
    let appName: String

    var body: some View {
        VStack {
            Text(appName)
                .font(.largeTitle)
                .padding()
```

```
                // Add more detailed information about the app here

                Spacer()
            }
            .navigationBarTitle(appName)
        }
    }
}

struct MainView: View {
    @State private var searchText = ""
    @State private var selectedApp: String? = nil

    private let adaptiveColumns = [
        GridItem(.adaptive(minimum: 300))
    ]

    var body: some View {
        NavigationView {
            List {
                Label("Discover", systemImage: "star")
                    .onTapGesture {
                        print("Discover")
                    }
                Label("Arcade", systemImage: "gamecontroller")
                    .onTapGesture {
                        selectedApp = "Arcade"
                    }
                Label("Create", systemImage: "paintbrush")
                    .onTapGesture {
                        selectedApp = "Create"
                    }
                Label("Categories", systemImage: "square.grid.3x3.
square")
                    .onTapGesture {
                        selectedApp = "Categories"
                    }
                Label("Favorites", systemImage: "heart")
                    .onTapGesture {
                        selectedApp = "Favorites"
                    }
                Label("Updates", systemImage: "square.and.arrow.down")
                    .onTapGesture {
                        selectedApp = "Updates"
```

```
                }
            }
            .searchable(text: $searchText)
            .onSubmit(of: .search) {
                print(searchText)
            }

            ScrollView {
                Color.clear

                Image("FireDEV Banner")
                    .resizable()
                    .padding(.horizontal)
                    .scaledToFit()

                LazyVGrid(columns: adaptiveColumns, spacing: 20) {
                    ForEach(0..<20) { index in
                        VStack(alignment: .leading) {
                            Image("FireDEV Logo")
                            Label("FireDEV", systemImage: "")
                                .font(.system(size: 36))
                                .onTapGesture {
                                    selectedApp = "FireDEV"
                                }
                        }
                    }
                }
            }
        }
        .sheet(item: $selectedApp) { app in
            AppDetailsView(appName: app)
        }
    }
}

struct MainView_Previews: PreviewProvider {
    static var previews: some View {
        MainView()
    }
}
```

In this modified code, I added a new `AppDetailsView`, which takes the selected app name as a parameter and displays more detailed information about the app. You can customize the content of this view based on your requirements.

I also added a new `@State` variable called `selectedApp` to track the selected app name. When a user taps on an app in the list or label, the corresponding app name is assigned to `selectedApp`, and `AppDetailsView` is presented as a sheet using `.sheet(item: $selectedApp)`. When the user dismisses the sheet, `selectedApp` is set back to `nil`.

In `AppDetailsView`, I simply display the app name for demonstration purposes. You can add more information and customize the layout as per your app's requirements.

With these modifications, when a user taps on an app, a new sheet will appear showing the enhanced page with more information about the selected app.

Summary

In this chapter, we successfully implemented the main body for the app store application. We started by analyzing the wireframes and breaking down each element into SwiftUI components. We then meticulously implemented the SwiftUI components to match the design from the wireframe. We implemented a scrollable stack with a highlight banner and app icons. We also looked at a few extra task implementations at the end of the chapter.

In the next chapter, we will begin working on our fourth and final application, which is the fitness companion app for Apple Watch. Our focus will be on analyzing the design and breaking it down to gain a better understanding of how we can implement this application on our next platform.

8

Watch Project – Fitness Companion Design

In the previous six chapters, we have created various applications for our Apple devices. These chapters taught us how to set up projects for the iPhone, iPad, and Mac. They also demonstrated the design differences between small and large displays. In this chapter, we will be designing a fitness companion application for the Apple Watch. Due to the small screen size of the watch, we will need to simplify the design. We will assess the requirements and discuss the design specifications before starting the coding process.

Firstly, we will assess the requirements needed for designing a fitness companion application for the Apple Watch. We will then move on to discussing the design specifications, which will give us a better understanding of what is required and how it will all fit together. This will be followed by the coding process, where we will build the fitness app over the course of these two chapters. This project will cover the foundations of SwiftUI components. We will discuss all this in the following sections:

- Understanding the design specifications
- Building the fitness app

In this chapter, we will gain a better understanding of the requirements and design of our application. The foundations we have established over the previous chapters in SwiftUI components, design, and Xcode navigation will serve as a strong starting point for the next chapter. Stay tuned for more progress as we continue to build upon these foundations.

Technical requirements

This chapter requires you to download Xcode version 14 or above from Apple's App Store.

To install it, just search for Xcode in the App Store, then select and download the latest version. Open Xcode and follow any additional installation instructions. Once Xcode has opened and launched, you're ready to go.

Version 14 of Xcode has the following features/requirements:

- Includes SDKs for iOS 16, iPadOS 16, macOS 12.3, tvOS 16, and watchOS 9.

- Supports on-device debugging in iOS 11 or later, tvOS 11 or later, and watchOS 4 or later.

- Requires a Mac running macOS Monterey 12.5 or later.

You can download the sample code from the following GitHub link: `https://github.com/PacktPublishing/Elevate-SwiftUI-Skills-by-Building-Projects`

In the next section, we will provide clarity on the specifications of our application's design and look at mockups of what the app will look like.

Understanding the Design Specifications

This section outlines the design specifications for our fitness companion application. Our goal is to implement features that will enhance the user experience. To achieve this, we have put ourselves in the user's shoes to determine how they will use the app. We have then broken down the process into individual steps to identify the necessary features. By doing this, we can ensure that our fitness app will be user-friendly and efficient.

In this section, we will look at the design specifications of our fitness companion application and describe the features we are going to implement. The best method for figuring out the features required is to put yourself in the user's shoes to determine how they will use the app and break it down into individual steps.

Our fitness application has been designed with several features to assist users in achieving their fitness goals. We are confident that our app will provide users with a seamless and effective fitness experience.

The features we would like our app to have are as follows:

- Current time
- Active/workout time
- BPM (beats per minute)
- Total calories burned
- Activity
- Views that can be swiped
- Starting a new workout
- Pausing a workout
- Ending a workout
- Locking a workout

- Goals
- Different exercises

After listing the ideal features, it is crucial to determine which ones are mandatory. Understanding the end use of our product is key. For me, the purpose of creating this fitness companion app is to provide a solid foundation for adding more advanced functionality later. We will not be implementing all the features in the preceding list, as it would be beneficial to try and implement them on your own as extra tasks to put the concepts you've learned into practice. Therefore, the following are the core features we will be implementing:

- Current time
- Active/workout time
- BPM (beats per minute)
- Total calories burned
- Activity
- Views that can be swiped
- Starting a new workout icon
- Pausing a workout icon
- Ending a workout icon
- Locking a workout icon

Once you have finished reading this chapter and the following one, you will be ready to tackle the remaining features on your own. The next section will outline the acceptance criteria for our application, providing you with the necessary guidelines to ensure its success.

Acceptance criteria for our app

In this section, we will outline the mandatory requirements for our application. These requirements are crucial for the end product and must be measurable. We need to ensure that these requirements are met to deliver a successful application. Let's get started:

- Current time – this will show the actual time as per your time zone.
- Active/workout time – this will be a live timer that shows the current workout time.
- BPM (beats per minute) – this label will be linked to a variable to show the user's heartbeats per minute.
- Total calories burned – this will display the calories burned and will be linked to the calories variable.

- Activity – this will be used to display the current activity; for example, running, swimming, yoga, and so on.

- Views that can be swiped – this will allow us to have two separate screens on a single page and expand it as our needs increase.

- Starting a new workout – a button composed of an image and text item will allow the user to start the workout.

- Pausing a workout icon – this button will also be composed of an image and text item that will be used to pause the workout.

- Ending a workout icon – another button, similar to the previous two, will be used for ending the workout.

- Locking a workout icon – finally, this button will be used for locking the workout.

To ensure the success of the application, it is crucial to develop test cases that measure the acceptance criteria. These test cases should simulate real-life scenarios and conditions in which the end user will use the application. By doing so, we can accurately measure the performance level that needs to be obtained for the application to be considered successful. Therefore, creating detailed test cases or scenarios is necessary to ensure the application meets the expected standards.

Wireframe for our app

Wireframing is an essential tool for designing layouts. It provides an overview of how the layout will look. The wireframe for the current activity in the fitness application is depicted in the following figure:

Figure 8.1 – Wireframe for our watch app

The following figure shows the wireframe for the view that allows you to start, stop, and pause activities:

Figure 8.2 – Activity button wireframe

We have now seen the wireframes for our fitness application. These wireframes will serve as the initial foundation for building the UI of our application.

In the next section, we will construct the interface for our application and verify that it matches the design we created in the wireframe. While we will follow the same process, there may be minor discrepancies. Our primary focus will be on the initial view in this chapter, with the second view being addressed in the subsequent chapter.

Building the Fitness App

We will now build the UI for the sidebar. First, let's create our project. Follow these steps:

1. Open Xcode and select **Create a new Xcode Project**:

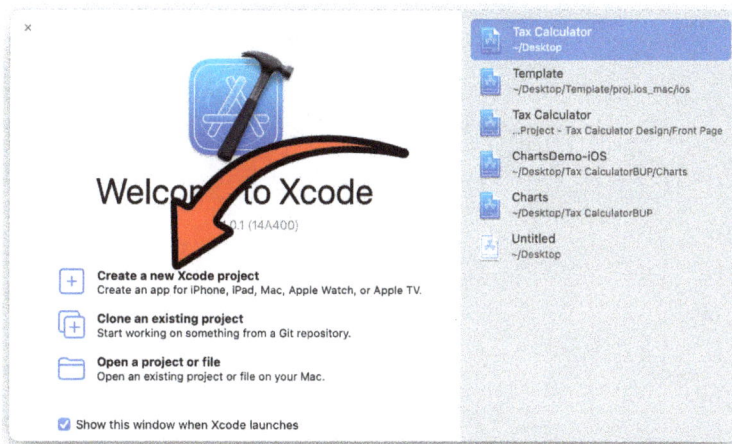

Figure 8.3 – Create a new Xcode project

2. Now we will choose the template for our application. As we are creating an Apple Watch application, we will select **WatchOS** from the top, then select **App**, and click **Next**:

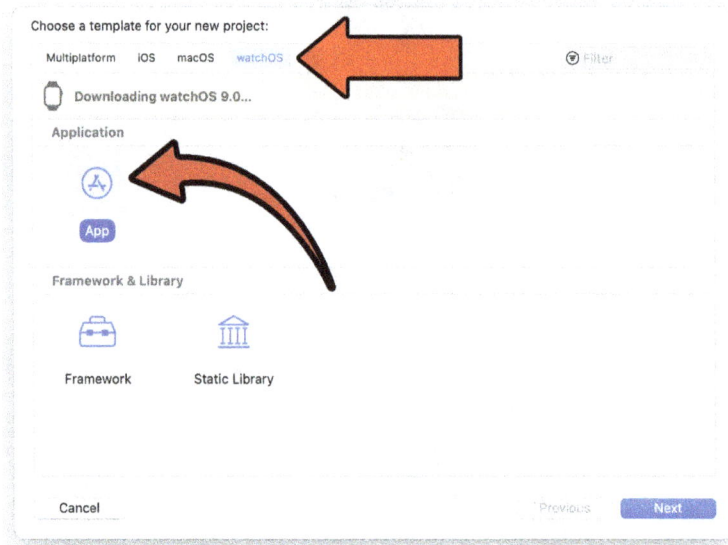

Figure 8.4 – Xcode project template selection

3. We will now choose the options for our project. Here, there is only one crucial thing to select/set. Make sure **Watch-only App** is selected:

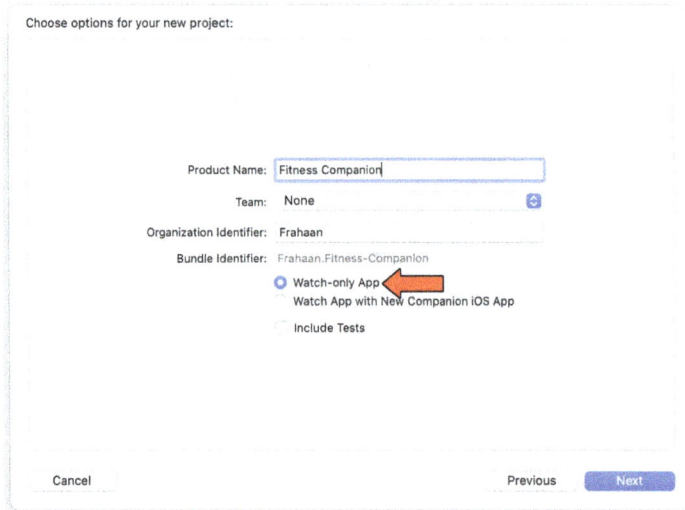

Figure 8.5 – Xcode project options

4. Once you press **Next**, you can choose where to create your project, as seen in the following figure:

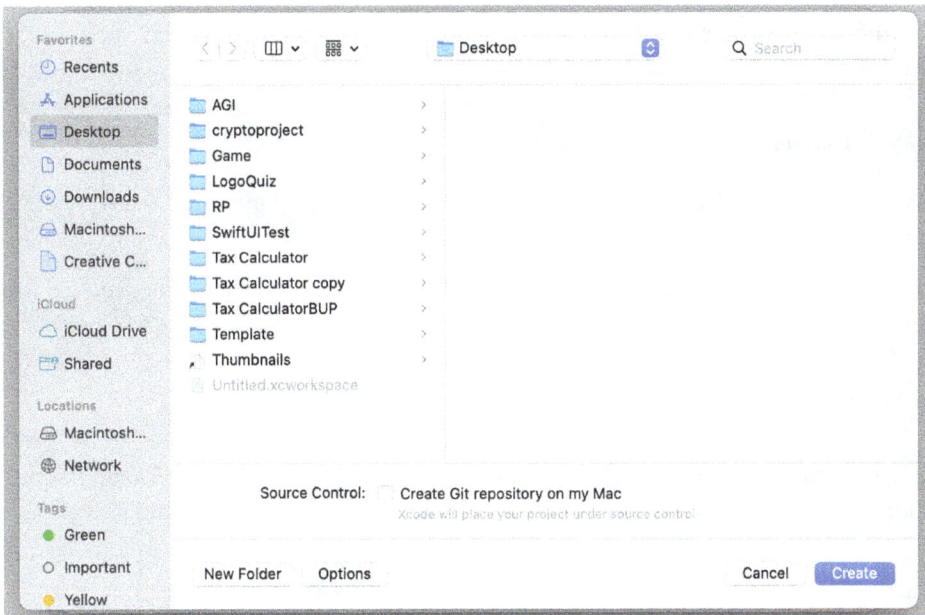

Figure 8.6 – Xcode project save directory

5.   Once you have found the location where you would like to create it, click **Create** in the bottom right. Xcode shows your project in all its glory, as seen in the following figure:

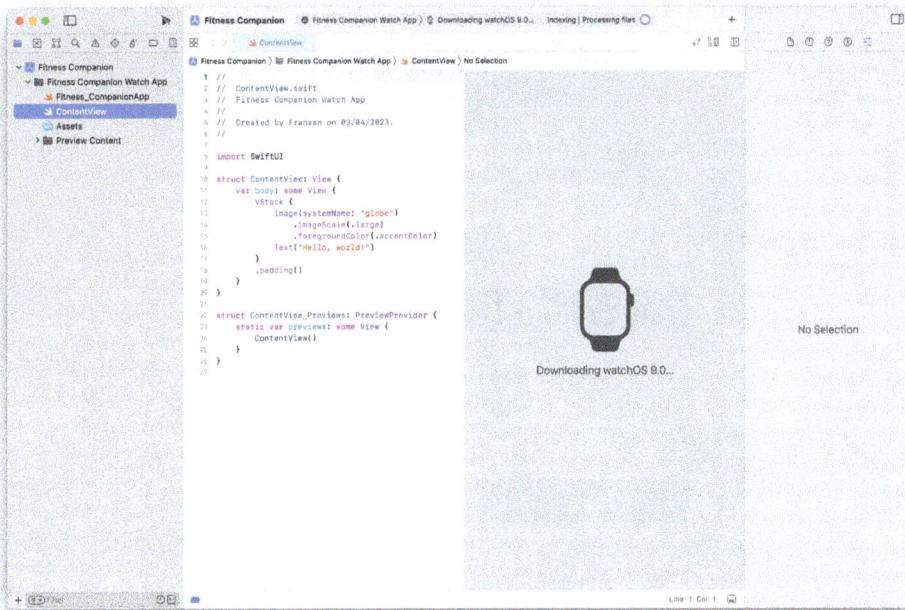

Figure 8.7 – New Xcode project overview

In this section, we set up our WatchOS project. Now that we are all set up, we will implement the interface for the first page of our fitness application.

Activity Details

In this section, we will implement the first page of the fitness application, which will represent the current activity details. As a reminder, refer to *Figure 8.1* to see what it looks like.

There are five main elements to the current activity screen. As a little task, see whether you can figure out what they are. Don't worry if you don't know the exact UI component names; we will look at these components in the following section.

Text

A text component displays a string of characters, numerals, or even icons, all of which can be used in conjunction with each other. For us, it will be used to display all five components as follows:

1. Current time:

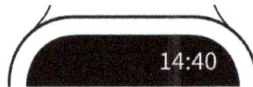

Figure 8.8 – Current time label

2. Activity running time:

Figure 8.9 – Activity time label

3. Beats per minute (BPM):

Figure 8.10 – BPM label

4. Calories burned:

Figure 8.11 – Calories label

5. Current activity:

Figure 8.12 – Current activity label

> **Important note**
>
> To rename views, please refer to the *Renaming views* section in the previous chapter to revisit the concept.

In this section, we analyzed the acceptance criteria along with the requirements of our fitness companion application. We also broke down the wireframes, enabling us to determine how the application works and is structured. We will utilize this knowledge going forward into the next section.

Implementing the Current Activity UI

In this section, we will use the newly set up project to start the coding of our fitness watch application. We will implement the current activity UI, which will display the information for the current activity.

As we have created a fresh project, the coding standards aren't in line with my personal preference. So, firstly, I will change the standards. Feel free to take a few moments to do the same.

If you run the newly created application as it is, you will notice that we already have the current time in the top right, as shown in the following screenshot:

Figure 8.13 – Default time

It's great for us, as the time is already in place by default. There may be scenarios where you want to remove the time, but as we do not, we can proceed. Now, we will move on to implementing the text items for the current activity. Implementing the remaining text items is actually very simple. Remove all the code within the VStack and add the following labels:

```
var body: some View
{
    VStack( )
    {
        Text( "00:10:44" )

        Text( "120 BPM" )

        Text( "110 Calories" )

        Text( "Running" )
    }
    .padding( )
}
```

This will result in the following layout:

Figure 8.14 – Text items added

Firstly, we will set the alignment of the VStack to left-aligned, like so:

```
VStack( alignment: .leading )
```

Though all the content is present, it's not dynamic or styled. First, we will make it dynamic. To do so, create five variables to store the following:

- Time counter – this will count up every second; it can be modified for more frequent counting.

- Timer string – this will use the time counter and convert it into 00:00:00 format.

- BPM – this will store the BPM number.

- Calories – this will store the number of calories burned during the workout session.

- Activity – this will inform the user which workout is active.

The code for this is as follows:

```
@State private var counter = 0
@State private var timerString = "00:00:00"
@State private var bpm = 120
@State private var calories = 110
@State private var activity = "Running"
```

Next, we will create a timer that runs every second:

```
let timer = Timer.publish(every: 1, on: .main, in: .common).
autoconnect()
```

Now, it is time to link these variables and the timer to the appropriate components. Firstly, add an onReceive event to the timer text component, like so:

```
Text( timerString )
    .onReceive( timer )
    { time in
        counter += 1

        let hours = counter / 3600
        let minutes = ( counter % 3600 ) / 60
        let seconds = counter % 3600 % 60

        timerString = String( hours ) + ":"
+ String( minutes ) + ":" +
String( seconds )
    }
```

Let's take a look at what we just did. The onReceive event takes the timer as the parameter, which is used to observe how often the published events are triggered by the timer. In each pass, we increment the counter by one, so the counter is the number of seconds elapsed. Then, we create constants for hours, minutes, and seconds. We do some simple mathematical calculations to figure out how many hours, minutes, and seconds the timer has been running. Finally, we will format our `timerString` to show `hours:minutes:seconds`. Running the application will show the following result:

Figure 8.15 – Dynamically linked variables

You may notice that the time is currently in the format of `0:0:0`, not `00:00:00`. Fixing this is super simple; we need to add a formatter onto each string to format it using two decimal places for the hours, minutes, and seconds. Update the `timerString` as follows:

```
timerString = String( format: "%02d", hours ) + ":" + String( format:
"%02d", minutes ) + ":" + String( format: "%02d", seconds )
```

Now, the application will appear as follows:

Figure 8.16 – Formatted timerString

It is looking better now. The BPM, calories, and current activity labels are all fine; we just need to modify the current activity time label. Three things need to be done: make it bigger, change the color to yellow, and add some padding below. Doing all of this is simple. Modify the current activity time text item, like so:

```
Text( timerString )
    .font( .title2 )
    .foregroundColor( Color.yellow )
    .padding( .bottom )
    .onReceive( timer ) …
```

Once we run the application, we will see we have completed the code for this chapter.

Figure 8.17 – Current activity time styled

In this section, we implemented the current activity screen for our fitness companion application. You learned that even though Apple WatchOS is the newest of the four SDKs and seems difficult, it is just the same and very simple to use. In the next chapter, we will implement a swipe view to add the activity button screen.

Summary

In this chapter, we covered the design of our fitness companion application. We looked at wireframes and broke each element down into SwiftUI components. We then implemented the SwiftUI components to match the design from the wireframes for the current activity screen. We also looked at the requirements and design specifications for building this application, then simplified it to the core features our app will provide.

Next, we designed our fitness companion app by creating wireframes and breaking down each element into SwiftUI components. We implemented these components to match the wireframe design for the current activity screen. We also reviewed the requirements and design specifications for building the app and simplified it to focus on the core features it will provide. There are always features that are nice to have but are inevitably cut from the first release, or what many call the **MVP (minimum viable product)**. This is effectively what we did. It is crucial to curb the scope to prevent it from becoming too large and out of hand.

In the next chapter, we'll look at implementing the activity button screen for our fitness companion application.

9

Watch Project – Fitness Companion UI

In this chapter, we will implement the activity button screen for the Fitness Companion project. In the previous chapter, we looked at the design of Fitness Companion and, more specifically, the **Current Activity** screen design. Then we broke the screen down into all the components required. We then implemented all the components using SwiftUI. At the end of the previous chapter, we only had a single screen that couldn't be swiped. The main section will be swipeable and present the user with a list of buttons for controlling the current activity. Then, we will analyze the activity button screen, break it down into all the components it is composed of, and implement all the components to provide a fitness app-like feel.

This chapter will be split into the following sections:

- Activity button screen overview
- Implementing the activity button screen
- Extra tasks

By the end of this chapter, you will have created a fitness companion application for WatchOS. This will serve as a template with a swipeable screen to show the user information about an activity. It will serve as a solid foundation for further expanding the fitness application or pivoting the project to something different while using the core structure we have implemented. As we reach the end of the chapter, I will provide exercises to implement more advanced functionality in the fitness companion app. This will be the fourth and last project in this book, providing you with a 360-degree view of iOS UI development using Swift.

Technical Requirements

This chapter requires you to download Xcode version 14 or above from Apple's App Store.

To install Xcode, just search for Xcode in the App Store and select and download the latest version. Open Xcode and follow any additional installation instructions. Once Xcode has opened and launched, you're ready to go.

Version 14 of Xcode has the following features/requirements:

- It includes SDKs for iOS 16, iPadOS 16, macOS 12.3, tvOS 16, and watchOS 9
- It supports on-device debugging in iOS 11 or later, tvOS 11 or later, and watchOS 4 or later
- You will require a Mac running macOS Monterey 12.5 or later

Download the sample code from the following GitHub link:

```
https://github.com/PacktPublishing/Elevate-SwiftUI-Skills-by-Building-
Projects
```

Activity button screen overview

In this section, we will take another look at the wireframe from *Chapter 8* and break it down into its components. The following figure showcases the activity button screen:

Figure 9.1 – Activity button screen

Before we code our application, we will break down the activity button screen into the elements that comprise it. As a little task, see whether you can figure out what these are. Don't worry if you don't know the exact UI component names; we will look at the components in the following section.

Image components

An Image component is one of the core components offered by SwiftUI. It allows you to display an image, which can be used to provide a visual representation or to aid a body of text. We will use it to display icons for buttons to control the current activity. The following figures show the icons from the application:

Figure 9.2 – Lock image

Figure 9.3 – New image

Figure 9.4 – End image

Figure 9.5 – Pause image

These images contain not only an icon but also a background. This is something that will be further explored later in this chapter.

Text components

We will be using text components to display button titles. Refer to *Chapter 2* for more information.

In the next section, we will implement the code for the activity button screen using the components we discussed in the previous sections.

Implementing the Activity Button Screen

In this section, we will implement our application's activity button screen and thus complete the fourth and final project in this book. Before we do this, we must implement a swipeable page system. The first page will contain the implementation from the previous chapter, and the second page will be the activity buttons. Naturally, you can use this to expand to as many pages as you require.

Swipeable Pages

In this section, we will implement our swipeable pages. Luckily for us, it is super simple to implement as many things as possible in SwiftUI. Simply enclose our current VStack in the MainView inside a TabView as demonstrated here:

```
TabView
{
    VStack( alignment: .leading )
    {
        Text( timerString )
            .font( .title2 )
            .foregroundColor( Color.yellow )
            .padding( .bottom )
            .onReceive( timer )
        { time in
            counter += 1

            let hours = counter / 3600
            let minutes = ( counter % 3600 ) / 60
            let seconds = counter % 3600 % 60

            timerString = String( format: "%02d", hours ) + ":"
+ String( format: "%02d", minutes ) + ":" +
String( format: "%02d", seconds )
        }
```

```
        Text( String( bpm ) + " BPM" )

        Text( String( calories ) + " Calories" )

        Text( activity )
    }
    .padding( )
}
```

In the preceding code, we implemented a `TabView`, which is used to create multiple pages in our fitness companion app.

Now, if you run it, it will look the same. However, if you try and swipe the screen, you will notice a bit of bounce. This is because there is only a single page. Now, let's add a dummy second page to help us test our new `TabView`. After the `VStack`, add a `Text` component, like so:

```
TabView
{
    VStack( alignment: .leading )
    {
        ...
    }
    .padding( )

    Text( "Second Page" )
}
```

Now we have implemented the second page in our `TabView`. If you try and swipe on the page, it will go to the next page.

Each view within the `TabView` is treated as an individual page. Honestly, it's as simple as that. Running our application will result in the following, which shows two dots at the bottom that indicate that there are two pages:

Figure 9.6 – First page

Swiping from right to left will show the second page, as follows:

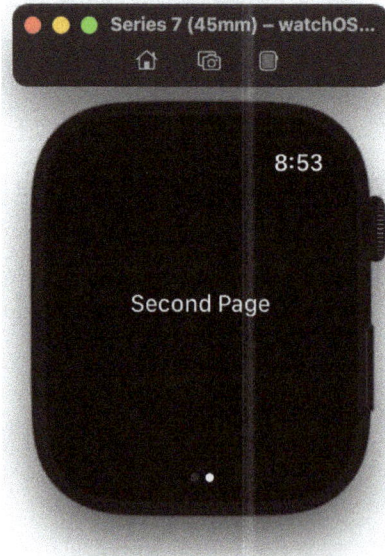

Figure 9.7 – Second page

In this section, we added an extra page using the `TabView` component. This allowed us to add another page that the user was able to traverse with a swipe gesture. In the next section, we will add the activity buttons.

Activity Buttons

In this section, we will implement the activity buttons on the second page of our `TabView`. We will be using custom colors for the background of each button and the icon itself.

Let's go ahead and create these custom colors:

1. Navigate to the **Assets** section within the **Project Navigator**:

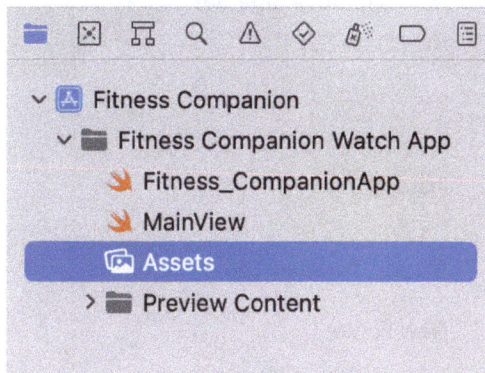

Figure 9.8 – Assets folder

2. In the **Assets** section, right-click on the empty space and select **New Color Set**:

☐ AccentColor
☐ AppIcon

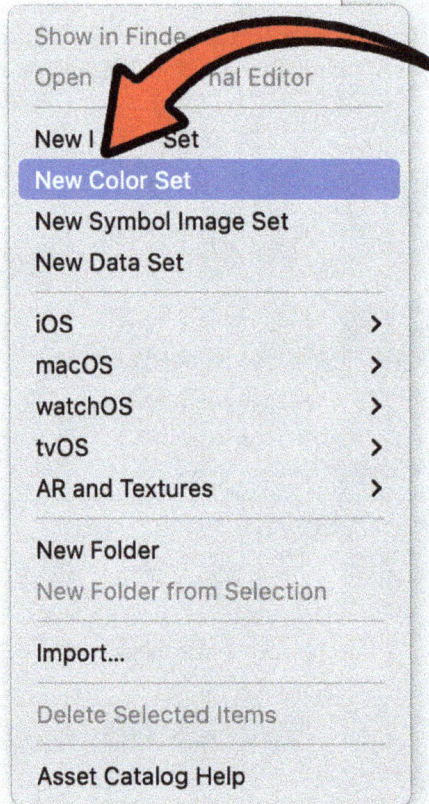

Show in Finde
Open   hal Editor

New I   Set
New Color Set
New Symbol Image Set
New Data Set

iOS   >
macOS   >
watchOS   >
tvOS   >
AR and Textures   >

New Folder
New Folder from Selection

Import...

Delete Selected Items

Asset Catalog Help

Figure 9.9 – New Color Set button

3. In **Attributes inspector**, set the name of the color:

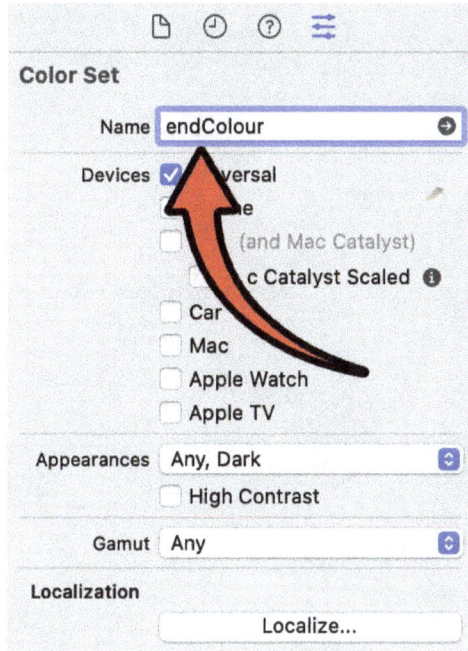

Figure 9.10 – New Color Set button

4. Select **Any Appearance** or **Dark** to set the color. This ensures that in all color modes, the desired color will be used:

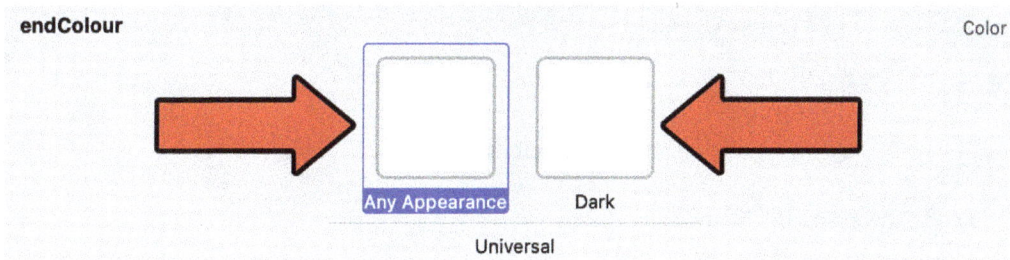

Figure 9.11 – Any Appearance

5. Now make sure **Content** is set to **sRGB** and **Input Method** is set to **8-bit Hexadecimal**. Then set the **Hex** value:

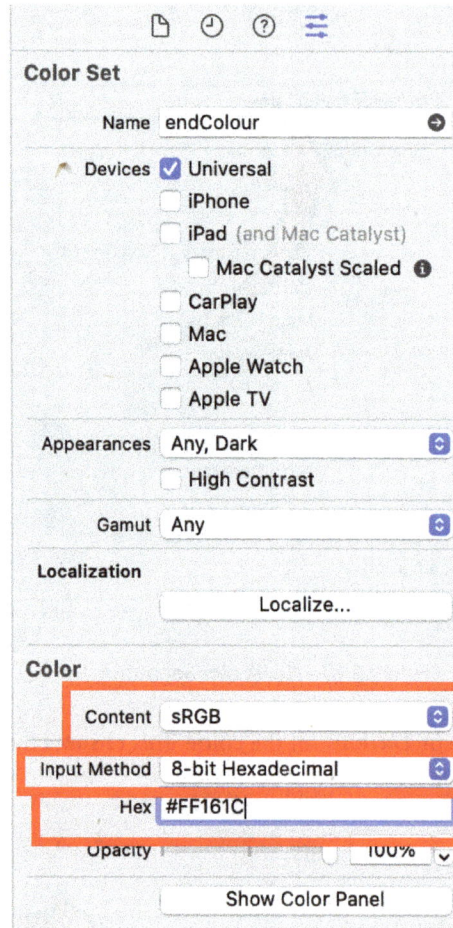

Figure 9.12 – Setting the color

6. Repeat these steps for all the colors listed:

- **End Button Colour (already added)**:

    - Name: `endColour`

    - Hex color value: `#FF161C`

- **End Background Button Colour**:

  - Name: `endColourBackground`
  - Hex color value: `#390B0C`

- **Lock Button Colour**:

  - Name: `lockColour`
  - Hex color value: `#06F5E7`

- **Lock Background Button Colour**:

  - Name: `lockColourBackground`
  - Hex color value: `#113330`

- **New Button Colour**:

  - Name: `newColour`
  - Hex color value: `#86FE01`

- **New Background Button Colour**:

  - Name: `newColourBackground`
  - Hex color value: `#1E3400`

- **Pause Button Colour**:

  - **Name**: `pauseColour`
  - Hex color value: `#BBA700`

- **Pause Background Button Colour**:

  - **Name**: `pauseColourBackground`
  - Hex color value: `#342F00`

7.   Once completed, the **Assets** screen should look as follows:

AccentColor
AppIcon
endColour
endColourBackground
lockColour
lockColourBackground
newColour
newColourBackground
pauseColour
pauseColourBackground

Figure 9.13 – Colors added

Even though the colors have been created, we cannot use them directly in our code. Let's fix this. Doing so is rather simple. In the `MainView`, we will extend the `Color` functionality to support our colors. Before doing so, I thought it is prudent to mention why I spelled color like so when referring to the functionality. This is the American spelling and is the one used within Swift. To extend the functionality, we must spell it this way, but as I am British, the spelling I personally use is *Colour*. Now that is cleared up. Let's extend the Swift color. Add the following code above the `MainView` struct:

```
extension Color
{
    static let lockColour = Color( "lockColour" )
    static let lockColourBackground = Color( "lockColourBackground" )

    static let newColour = Color( "newColour" )
    static let newColourBackground = Color( "newColourBackground" )

    static let endColour = Color( "endColour" )
    static let endColourBackground = Color( "endColourBackground" )

    static let pauseColour = Color( "pauseColour" )
    static let pauseColourBackground = Color( "pauseColourBackground"
    )
}
```

Though we have added it to the `MainView`, extending the Swift color allows us to use it anywhere in our project. It's also worth mentioning that this means only in our project and doesn't extend beyond the scope of our project into other projects. Our project is small, so it's perfectly fine putting it inside the `MainView`. However, it is common practice to put extensions like this in a specific file. If there are a lot of color extensions, then they could have their own Color file. This is beyond the scope of this project.

Here's a quick overview of the preceding code: we extended the Color using static variables of the names we set previously in the asset.

> **Important note**
> The variable names do not need to be the same as the color name. But it is good practice to keep them the same. It makes them easier to maintain.

Each of the buttons is made up of three components:

- Background
- Icon
- Text

For the background, we will use a `Rectangle` component. More specifically, we will use the `RoundedRectangle` component as it allows us to set a corner radius. Feel free to change the design and use a `Rectangle` or any other shape. For the icon, we will use the `Image` component and use a built-in icon. Feel free to use your own image or look at SF Symbols, as discussed earlier, in *Chapter 6, Implementing the Sidebar*, to see all built-in icons. The text is the simplest of all the components and will use a basic `Text` component.

The most difficult part of the buttons is the background and icon. This is because they are on top of each other. The text is placed below the image, making it easy to add. We will initially concentrate on getting the background and icon coded. We will use a `ZStack` to place the icon on top of the background. Replace the second-page dummy `Text` component with the following code:

```
ZStack
{
    RoundedRectangle( cornerRadius: 18, style:
.continuous )
}
```

We have created a `RoundedRectangle` with a corner radius of 18. Feel free to increase the number to get more rounded corners, or to lower it. Setting the style to `.continuous` makes the corners appear smoother, which is always a good thing. Let's see what this produced:

Figure 9.14 – Simple rounded rectangle

Right now, it's not looking anything like the button in the figures shown at the start of this chapter. There are just two things missing – the background color and making the size smaller. We will use the lockColourBackground color we created previously and set the size to a width of 70 and a height of 64:

```
ZStack
{
    RoundedRectangle( cornerRadius: 18, style:
.continuous )
        .foregroundColor( .lockColourBackground )
        .frame( width: 70, height: 64 )
}
```

The following figure shows the image background:

Figure 9.15 – Rounded color styled

The background is finally looking more like our design. The next step is to add the icon inside of the rectangle. Doing so is simple. Add an image with the icon after the `RoundedRectangle` component:

```
ZStack
{
    RoundedRectangle( cornerRadius: 18, style:
.continuous )
        .foregroundColor( .lockColourBackground )
        .frame( width: 70, height: 64 )

    Image( systemName: "drop.fill" )
}
```

Running the application now will produce the following:

Figure 9.16 – Teardrop icon in background

We used **SF Symbols** to obtain the teardrop icon. Feel free to use any icon as you see fit, or even your own image. We need to change the icon style. There are two main aspects to update, the color and size. Update the image as follows:

```
Image( systemName: "drop.fill" )
    .resizable( )
    .foregroundColor( .lockColour )
    .aspectRatio( contentMode: .fit )
    .frame( width: 16 )
```

We first set it to `resizable`, which allows us to change the size. Next, we set the color using one of the colors we created earlier. Next, we ensure the aspect ratio is set to fit, which allows us to resize the image without distorting it. Finally, we set the size because we have a locked aspect ratio. Setting the width automatically sets the height accordingly. Running the application produces the result that follows:

Figure 9.17 – Lock button

Next, we will add text to our button. The text isn't inside of the icon or even the rectangle but sits beneath it. But we still want it to be clickable, so we will wrap all of the button contents inside a VStack and also add the Text component, like so:

```
VStack
{
    ZStack
    {
        RoundedRectangle( cornerRadius: 18, style:
.continuous )
            .foregroundColor( .lockColourBackground )
            .frame( width: 70, height: 64 )

        Image( systemName: "drop.fill" )
            .resizable( )
            .foregroundColor( .lockColour )
            .aspectRatio( contentMode: .fit )
            .frame( width: 16 )
    }
    Text( "Lock" )
}
```

The reason for wrapping the button in a VStack is two-fold:

- We want all of it to be clickable (to be implemented next).

- As there will be multiple buttons, the VStack is technically the button without the icon, rectangle, or text components. Pretty cool!

Let's make the VStack clickable, and then we will take a look at the result. First, we need a function for it to call. We could use an inline function, but we will create a dedicated function. This provides a nice abstraction in our code base. Before the body, add the following code:

```
func Lock( )
{ print( "Lock button is pressed" ); }

var body: some View
{
    ...
}
```

The function is a simple one. When clicked, a message is logged to the terminal. Now update the VStack with an onTapGesture function:

```
VStack
{
    ZStack
    {
        RoundedRectangle( cornerRadius: 18, style:
.continuous )
            .foregroundColor( .lockColourBackground )
            .frame( width: 70, height: 64 )

        Image( systemName: "drop.fill" )
            .resizable( )
            .foregroundColor( .lockColour )
            .aspectRatio( contentMode: .fit )
            .frame( width: 16 )
    }

    Text( "Lock" )
}.onTapGesture { Lock( ) }
```

That was a lot. Let's run our application and see the result:

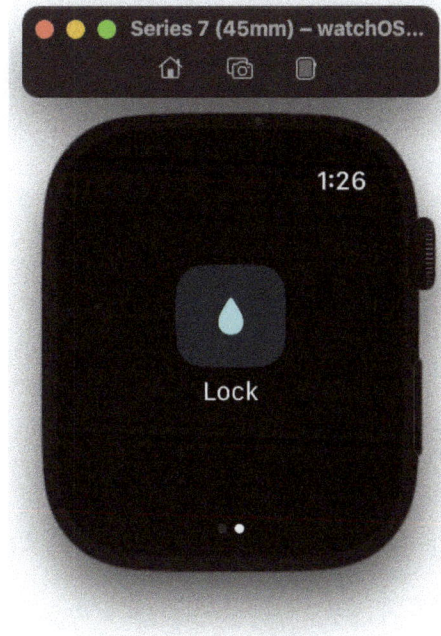

Figure 9.18 – Finished lock button

Feel free to click the button. It will log a message. Before wrapping up this project and implementing the remaining buttons, here is the code so far:

```
//
//  ContentView.swift
//  Fitness Companion Watch App
//
//  Created by Frahaan on 03/04/2023.
//

import SwiftUI

extension Color
{
    static let lockColour = Color( "lockColour" )
    static let lockColourBackground = Color( "lockColourBackground" )

    static let newColour = Color( "newColour" )
    static let newColourBackground = Color( "newColourBackground" )
```

```
    static let endColour = Color( "endColour" )
    static let endColourBackground = Color( "endColourBackground" )

    static let pauseColour = Color( "pauseColour" )
    static let pauseColourBackground = Color( "pauseColourBackground"
)
}

struct MainView: View
{
    @State private var counter = 0

    @State private var timerString = "00:00:00"
    @State private var bpm = 120
    @State private var calories = 110
    @State private var activity = "Running"

    let timer = Timer.publish( every: 1, on: .main, in: .common
).autoconnect( )

    func Lock( )
    { print( "Lock button is pressed" ); }

    var body: some View
    {

        TabView
        {
            VStack( alignment: .leading )
            {
                Text( timerString )
                    .font( .title2 )
                    .foregroundColor( Color.yellow )
                    .padding( .bottom )
                    .onReceive( timer )
                { time in
                    counter += 1

                    let hours = counter / 3600
                    let minutes = ( counter % 3600 ) / 60
                    let seconds = counter % 3600 % 60
```

```
                            timerString = String( format: "%02d", hours )
+ ":" + String( format: "%02d", minutes ) + ":" + String( format:
"%02d", seconds )
                    }

                Text( String( bpm ) + " BPM" )

                Text( String( calories ) + " Calories" )

                Text( activity )
            }
            .padding( )

            VStack
            {
                ZStack
                {
                    RoundedRectangle( cornerRadius: 18, style:
.continuous )
                        .foregroundColor( .lockColourBackground )
                        .frame( width: 70, height: 64 )

                    Image( systemName: "drop.fill" )
                        .resizable( )
                        .foregroundColor( .lockColour )
                        .aspectRatio( contentMode: .fit )
                        .frame( width: 16 )
                }

                Text( "Lock" )
            }.onTapGesture { Lock( ) }
        }

    }
}

struct MainView_Previews: PreviewProvider
{
    static var previews: some View
    {
        MainView( )
    }
}
```

We have almost completed the chapter; the only thing left is to add the remaining buttons. First, let's add the remaining function callbacks above the body, like so:

```
func Lock( )
{ print( "Lock button is pressed" ); }

func New( )
{ print( "New button is pressed" ); }

func End( )
{ print( "End button is pressed" ); }

func Pause( )
{ print( "Pause button is pressed" ); }

var body: some View
{

    . . .

}
```

Now that we have implemented the callbacks, we are going to implement the buttons. Each button itself is actually simple, as it is the same as the lock button but with the following changes:

- RoundedRectangle's foregroundColor

- Image's icon

- Image's foregroundColor

- Text

- The onTapGesture callback

Merely duplicating the VStack code would effectively be duplicating the button. We would have the following results:

Figure 9.19 – Extra buttons added

If you can't see a difference, I don't blame you. It's hard to see what's actually happened, but if you look at the bottom of the screen, there are five dots, indicating that there are five pages now. Remember, when implementing the second page, we stated that each component in the root would be its own page. So, we want to group all these buttons together by putting them in a 2x2 grid. To organize components on the same line, we can use an `HStack`. We only need two on a single line, so first, we will enclose the first two buttons, the `VStacks` in an `HStack`. This will result in three pages, the first being the one implemented in the previous chapter and two for each `HStack`, which isn't what we want. One small change and it will be fixed. Can you guess what it is? Just enclose both `HStacks` in a single `VStack`. This will put them on top of each other, thus rendering a grid. The code for this is as follows:

```
VStack
{
    HStack
    {
        VStack
        {
            ZStack
            {
                RoundedRectangle( cornerRadius: 18, style: .continuous
)
                    .foregroundColor( .lockColourBackground )
```

```
                    .frame( width: 70, height: 64 )

            Image( systemName: "drop.fill" )
                .resizable( )
                .foregroundColor( .lockColour )
                .aspectRatio( 1.0, contentMode: .fit )
                .frame( width: 32 )
        }


        Text( "Lock" )
    }.onTapGesture { Lock( ) }

    VStack
    {
        ZStack
        {
            RoundedRectangle( cornerRadius: 18, style: .continuous
)
                .foregroundColor( .newColourBackground )
                .frame( width: 70, height: 64 )

            Image( systemName: "plus" )
                .resizable( )
                .foregroundColor( .newColour )
                .aspectRatio( 1.0, contentMode: .fit )
                .frame( width: 32 )
        }


        Text( "New" )
    }.onTapGesture { New( ) }
}

HStack
{
    VStack
    {
        ZStack
        {
```

```
)                   RoundedRectangle( cornerRadius: 18, style: .continuous

                        .foregroundColor( .endColourBackground )
                        .frame( width: 70, height: 64 )

                    Image( systemName: "xmark" )
                        .resizable( )
                        .foregroundColor( .endColour )
                        .aspectRatio( 1.0, contentMode: .fit )
                        .frame( width: 32 )
                }


            Text( "End" )
        }.onTapGesture { End( ) }

        VStack
        {
            ZStack
            {
)                   RoundedRectangle( cornerRadius: 18, style: .continuous

                        .foregroundColor( .pauseColourBackground )
                        .frame( width: 70, height: 64 )

                    Image( systemName: "pause" )
                        .resizable( )
                        .foregroundColor( .pauseColour )
                        .aspectRatio( 1.0, contentMode: .fit )
                        .frame( width: 32 )
                }


            Text( "Pause" )
        }.onTapGesture { Pause( ) }
    }
}
```

It's time to run our app:

Figure 9.20 – Button grid system

You might be ready to finish this chapter and call it a day. Please bear with me – the top row of the grid is too close to the time for my liking. Let's add top **padding** to the outermost `VStack` we just implemented:

```
.padding( .top, 20.0 )
```

This is it, the final build and run of our application – drum roll, please!

Figure 9.21 – Button grid system

We are now at the end of our project, and our application looks amazing. Before we summarize, feel free to visit the GitHub repository to double-check your code base: `https://github.com/PacktPublishing/Elevate-SwiftUI-Skills-by-Building-Projects`.

In this section, we added the activity buttons. We did this by implementing another page in our fitness app. We leveraged a grid-based system for laying out our buttons. We used a variety of core components combined with stacks to organize them. In the following section, we will summarize this chapter, and ultimately this book. But first, we will look at some code to help you with the extra tasks.

Different Exercises

To add different exercises to the fitness companion app, you can modify the `MainView` by introducing a new data structure to store exercise information and update the UI accordingly. Here's an example of how you can make these changes:

```swift
import SwiftUI

extension Color {
    static let lockColour = Color("lockColour")
    static let lockColourBackground = Color("lockColourBackground")

    static let newColour = Color("newColour")
    static let newColourBackground = Color("newColourBackground")

    static let endColour = Color("endColour")
    static let endColourBackground = Color("endColourBackground")

    static let pauseColour = Color("pauseColour")
    static let pauseColourBackground = Color("pauseColourBackground")
}

struct Exercise {
    let name: String
    let image: String
}

struct MainView: View {
    @State private var counter = 0

    @State private var timerString = "00:00:00"
    @State private var bpm = 120
    @State private var calories = 110
    @State private var activity = "Running"
```

```swift
    let timer = Timer.publish(every: 1, on: .main, in: .common).
autoconnect()

    let exercises = [
        Exercise(name: "Running", image: "person.running"),
        Exercise(name: "Cycling", image: "bicycle"),
        Exercise(name: "Swimming", image: "figure.walk"),
        // Add more exercises here
    ]

    func Lock() {
        print("Lock button is pressed")
    }

    func New() {
        print("New button is pressed")
    }

    func End() {
        print("End button is pressed")
    }

    func Pause() {
        print("Pause button is pressed")
    }

    var body: some View {
        TabView {
            VStack(alignment: .leading) {
                Text(timerString)
                    .font(.title2)
                    .foregroundColor(Color.yellow)
                    .padding(.bottom)
                    .onReceive(timer) { time in
                        counter += 1

                        let hours = counter / 3600
                        let minutes = (counter % 3600) / 60
                        let seconds = counter % 3600 % 60

                        timerString = String(format: "%02d", hours) +
":" + String(format: "%02d", minutes) + ":" + String(format: "%02d",
seconds)
                    }
```

```
                    Text(String(bpm) + " BPM")

                    Text(String(calories) + " Calories")

                    Text(activity)
                }
                .padding()

            VStack {
                HStack {
                    ForEach(exercises, id: \.name) { exercise in
                        VStack {
                            ZStack {
                                RoundedRectangle(cornerRadius: 18,
style: .continuous)
                                    .foregroundColor(.
newColourBackground)
                                    .frame(width: 70, height: 64)

                                Image(systemName: exercise.image)
                                    .resizable()
                                    .foregroundColor(.newColour)
                                    .aspectRatio(contentMode: .fit)
                                    .frame(width: 32)
                            }

                            Text(exercise.name)
                        }
                        .onTapGesture {
                            activity = exercise.name
                        }
                    }
                }

                HStack {
                    VStack {
                        ZStack {
                            RoundedRectangle(cornerRadius: 18, style:
.continuous)
                                .foregroundColor(.
lockColourBackground)
                                .frame(width: 70, height: 64)
```

```
                              Image(systemName: "lock.fill")
                                  .resizable()
                                  .foregroundColor(.lockColour)
                                  .aspectRatio(contentMode: .fit)
                                  .frame(width: 32

)

                         }


                  Text("Lock")
              }.onTapGesture { Lock() }

              VStack {
                  ZStack {
                      RoundedRectangle(cornerRadius: 18, style:
.continuous)
                              .foregroundColor(.endColourBackground)
                              .frame(width: 70, height: 64)

                          Image(systemName: "xmark")
                              .resizable()
                              .foregroundColor(.endColour)
                              .aspectRatio(contentMode: .fit)
                              .frame(width: 32)
                  }


                  Text("End")
              }.onTapGesture { End() }

              VStack {
                  ZStack {
                      RoundedRectangle(cornerRadius: 18, style:
.continuous)
                              .foregroundColor(.
pauseColourBackground)
                              .frame(width: 70, height: 64)

                          Image(systemName: "pause")
                              .resizable()
                              .foregroundColor(.pauseColour)
                              .aspectRatio(contentMode: .fit)
                              .frame(width: 32)
```

```
                        }


                     Text("Pause")
                }.onTapGesture { Pause() }
            }
        }
        .padding(.top, 20.0)
    }
  }
}


struct MainView_Previews: PreviewProvider {
    static var previews: some View {
        MainView()
    }
}
```

In this modified code, we have introduced a new `Exercise` struct that stores the name and image name of each exercise. You can add more exercises to the `exercises` array by creating new `Exercise` instances.

In the view, we have used a `ForEach` loop to iterate over the exercises and display them dynamically. Each exercise is represented by a `VStack` containing an image and a text label. When an exercise is tapped, the activity state is updated with the selected exercise's name.

You can customize the exercise images and add more properties to the `Exercise` struct based on your specific requirements.

Active timer

To add functionality for an active timer that can be started, stopped, and paused, you can modify the `MainView` by introducing additional state variables and actions. Here's an example of how you can make these changes:

```
import SwiftUI

struct Exercise {
    let name: String
    let image: String
}

struct MainView: View {
    @State private var counter = 0
```

```
    @State private var isTimerRunning = false
    @State private var isTimerPaused = false

    @State private var timerString = "00:00:00"
    @State private var bpm = 120
    @State private var calories = 110
    @State private var activity = "Running"

    let timer = Timer.publish(every: 1, on: .main, in: .common).
autoconnect()

    let exercises = [
        Exercise(name: "Running", image: "person.running"),
        Exercise(name: "Cycling", image: "bicycle"),
        Exercise(name: "Swimming", image: "figure.walk"),
        // Add more exercises here
    ]

    func lock() {
        print("Lock button is pressed")
    }

    func startTimer() {
        isTimerRunning = true
        isTimerPaused = false
    }

    func pauseTimer() {
        isTimerRunning = false
        isTimerPaused = true
    }

    func stopTimer() {
        isTimerRunning = false
        isTimerPaused = false
        counter = 0
        timerString = "00:00:00"
    }

    var body: some View {
        TabView {
            VStack(alignment: .leading) {
                Text(timerString)
                    .font(.title2)
```

```
                    .foregroundColor(Color.yellow)
                    .padding(.bottom)
                    .onReceive(timer) { time in
                        if isTimerRunning && !isTimerPaused {
                            counter += 1

                            let hours = counter / 3600
                            let minutes = (counter % 3600) / 60
                            let seconds = counter % 3600 % 60

                            timerString = String(format: "%02d",
hours) + ":" + String(format: "%02d", minutes) + ":" + String(format:
"%02d", seconds)
                        }
                    }

                Text(String(bpm) + " BPM")

                Text(String(calories) + " Calories")

                Text(activity)
            }
            .padding()

            VStack {
                HStack {
                    ForEach(exercises, id: \.name) { exercise in
                        VStack {
                            ZStack {
                                RoundedRectangle(cornerRadius: 18,
style: .continuous)
                                    .foregroundColor(.
newColourBackground)
                                    .frame(width: 70, height: 64)

                                Image(systemName: exercise.image)
                                    .resizable()
                                    .foregroundColor(.newColour)
                                    .aspectRatio(contentMode: .fit)
                                    .frame(width: 32)
                            }

                            Text(exercise.name)
                        }
                        .onTapGesture {
```

```
                                    activity = exercise.name
                        }
                    }
                }

            HStack {
                VStack {
                    ZStack {
                        RoundedRectangle(cornerRadius: 18, style:
.continuous)
                            .foregroundColor(.
lockColourBackground)
                            .frame(width: 70, height: 64)

                        Image(systemName: "lock.fill")
                            .resizable()
                            .foregroundColor(.lockColour)
                            .aspectRatio(contentMode: .fit)
                            .frame(width: 32)
                    }

                    Text("Lock")
                }
                .onTapGesture { lock() }

                VStack {
                    if isTimerRunning

    {
                        Button(action: pauseTimer) {
                            ZStack {
                                RoundedRectangle(cornerRadius: 18,
style: .continuous)
                                    .foregroundColor(.
pauseColourBackground)
                                    .frame(width: 70, height: 64)

                                Image(systemName: "pause")
                                    .resizable()
                                    .foregroundColor(.pauseColour)
                                    .aspectRatio(contentMode:
.fit)
                                    .frame(width: 32)
                            }
                        }
```

```
                                    .buttonStyle(PlainButtonStyle())
                        } else {
                            Button(action: startTimer) {
                                ZStack {
                                    RoundedRectangle(cornerRadius: 18,
style: .continuous)
                                        .foregroundColor(.
pauseColourBackground)

                                        .frame(width: 70, height: 64)

                                    Image(systemName: "play.fill")
                                        .resizable()
                                        .foregroundColor(.pauseColour)
                                        .aspectRatio(contentMode:
.fit)

                                        .frame(width: 32)
                                }
                            }
                            .buttonStyle(PlainButtonStyle())
                        }

                        Text(isTimerRunning ? "Pause" : "Start")
                    }

                    VStack {
                        ZStack {
                            RoundedRectangle(cornerRadius: 18, style:
.continuous)
                                .foregroundColor(.endColourBackground)
                                .frame(width: 70, height: 64)

                            Image(systemName: "xmark")
                                .resizable()
                                .foregroundColor(.endColour)
                                .aspectRatio(contentMode: .fit)
                                .frame(width: 32)
                        }

                        Text("End")
                    }
                    .onTapGesture { stopTimer() }
                }
            }
            .padding(.top, 20.0)
```

```
            }
        }
    }
}

struct MainView_Previews: PreviewProvider {
    static var previews: some View {
        MainView()
    }
}
```

In this modified code, we have made the following changes:

- We introduced `isTimerRunning` and `isTimerPaused` state variables to track the timer's state.

- We added `startTimer()`, `pauseTimer()`, and `stopTimer()` actions to handle starting, pausing, and stopping the timer, respectively.

- We modified the "Start/Pause" button to toggle between the **Start** and **Pause** states based on the `isTimerRunning` state.

- We updated the timer's `onReceive` closure to only increment the counter and update the timer string when the timer is running and not paused.

- We added functionality to reset the counter and timer string when the **End** button is pressed.

With these changes, you can now start, pause, and stop the timer in your fitness companion app.

Summary

In this chapter, we successfully added the activity button screen to our Fitness Companion application. We began by analyzing the wireframe and breaking down each element into SwiftUI components. From there, we implemented the components to match the design from the wireframes. Through this process, we gained a deeper understanding of how to combine core SwiftUI components using stacks to create complex buttons. We finally looked at a few implementations for the extra tasks as well.

I want to express my gratitude for taking the time to read this book. There were moments when I questioned the purpose of writing yet another programming book, but ultimately, it was a worthwhile endeavor. I sincerely hope that you were able to gain something from it. If you have any questions or would like to contact me directly, please feel free to use any of the platforms listed here:

- Twitter: https://twitter.com/SonarSystems

- Email: support@sonarsystems.co.uk

- Discord: https://discord.gg/7e78FxrgqH

- FireDEV Podcast: https://open.spotify.com/show/387RiHksQE33KYHTitFXhg

Index

‹packt›

Other Books You May Enjoy

If you enjoyed this book, you may be interested in these other books by Packt:

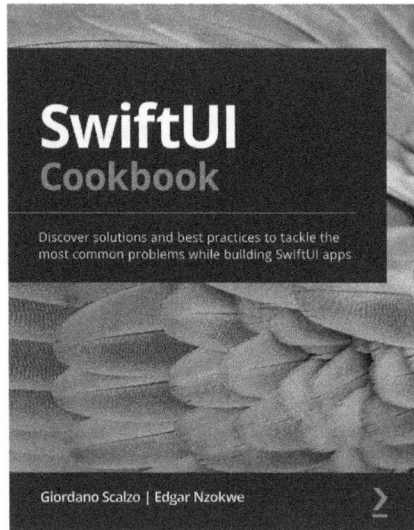

SwiftUI Cookbook

Giordano Scalzo | Edgar Nzokwe

ISBN: 978-1-83898-186-0

- Explore various layout presentations in SwiftUI such as HStack, VStack, LazyHStack, and LazyVGrid.
- Create a cross-platform app for iOS, macOS, and watchOS.
- Get up to speed with drawings in SwiftUI using built-in shapes, custom paths, and polygons.
- Discover modern animation and transition techniques in SwiftUI.
- Add user authentication using Firebase and Sign in with Apple.

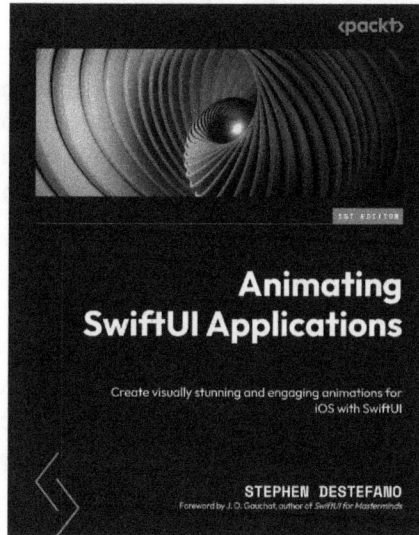

Animating SwiftUI Applications

Stephen DeStefano

ISBN: 978-1-80323-266-9

- Understand the fundamentals of SwiftUI and declarative programming.
- Master animation concepts like state variables and time curves
- Explore animation properties like hueRotation, opacity, and scale.
- Create animations using physics, gravity, collision, and more.
- Use the Geometry Reader to align views across various platforms Combine different animations for more dynamic effects.
- Add audio to your animations for an interactive experience.

Packt is searching for authors like you

If you're interested in becoming an author for Packt, please visit authors.packtpub.com and apply today. We have worked with thousands of developers and tech professionals, just like you, to help them share their insight with the global tech community. You can make a general application, apply for a specific hot topic that we are recruiting an author for, or submit your own idea.

Share Your Thoughts

Now you've finished *Elevate SwiftUI Skills by Building Projects*, we'd love to hear your thoughts! Scan the QR code below to go straight to the Amazon review page for this book and share your feedback or leave a review on the site that you purchased it from.

https://packt.link/r/1-803-24207-8

Your review is important to us and the tech community and will help us make sure we're delivering excellent quality content.

Download a free PDF copy of this book

Thanks for purchasing this book!

Do you like to read on the go but are unable to carry your print books everywhere? Is your eBook purchase not compatible with the device of your choice?

Don't worry, now with every Packt book you get a DRM-free PDF version of that book at no cost.

Read anywhere, any place, on any device. Search, copy, and paste code from your favorite technical books directly into your application.

The perks don't stop there, you can get exclusive access to discounts, newsletters, and great free content in your inbox daily

Follow these simple steps to get the benefits:

1. Scan the QR code or visit the link below

https://packt.link/free-ebook/9781803242071

1. Submit your proof of purchase
2. That's it! We'll send your free PDF and other benefits to your email directly

www.ingramcontent.com/pod-product-compliance
Lightning Source LLC
Chambersburg PA
CBHW080523220326
41599CB00032B/6180